Plumber's
Field Manual

A Selection of McGraw-Hill Books by R. Dodge Woodson

Journeyman Plumber's Licensing Exam Guide
Master Plumber's Licensing Exam Guide
Plumber's Quick Reference Manual
Plumber's Troubleshooting Guide
The National Plumbing Codes Handbook
The Plumbing Apprentice Handbook

Plumber's Field Manual

R. Dodge Woodson

McGraw-Hill

New York   San Francisco   Washington, D.C.   Auckland   Bogotá
Caracas   Lisbon   London   Madrid   Mexico City   Milan
Montreal   New Delhi   San Juan   Singapore
Sydney   Tokyo   Toronto

Library of Congress Cataloging-in-Publication Data

Woodson, R. Dodge (Roger Dodge)
    Plumber's field manual / R. Dodge Woodson.
        p.   cm.
    ISBN 0-07-071779-6
    1. Plumbing—Handbooks, manuals, etc.   I. Title.
  TH6125.W562   1996
  696'.1—dc20   96-31714
                                    CIP

McGraw-Hill

A Division of The McGraw·Hill Companies

1 2 3 4 5 6 7 8 9 0 DOC/DOC 9 0 1 0 9 8 7 6

ISBN 0-07-071779-6

*The sponsoring editor for this book was Larry Hager, the
editing supervisor was Caroline R. Levine, and the
production supervisor was Donald F. Schmidt. This book
was set in Century Schoolbook by Terry Leaden of
McGraw-Hill's Professional Book Group composition unit.*

Printed and bound by R. R. Donnelley & Sons Company.

This book is printed in recycled, acid-free paper containing a minimum of 50% recycled, de-inked fiber.

*Afton, Adam, and Kimberley are the
people who support me, and I dedicate
this book to them.*

Contents

Preface   xiii

Chapter 1. Installing Water Distribution Systems   1

Potable Water   2
Choosing the Type of Pipe and Fittings to Be Used   5
CPVC Pipe   6
Polybutylene Pipe   7
Copper Pipe   9
The Pipe Decision   10
Sizing   12
Installing the Potable Water System   16
Valves   17
Backflow Protection   18
A Manifold System   27
Tips for Underground Water Pipes   28

Chapter 2. Troubleshooting and Repairing Water Distribution Systems   29

Low Water Pressure   30
Excessive Water Pressure   34
Water Hammer   34
Leaks in Copper Piping   40
PB Leaks   46
Polyethylene Leaks   46
Chlorinated Polyvinyl Chloride Leaks   47
Galvanized Steel and Brass   48

Compression Leaks   50
Frozen Pipes   50

Chapter 3. Installing Drain-Waste-and-Vent Systems   53

Underground Plumbing   54
Aboveground Rough-ins   65

Chapter 4. Troubleshooting and Repairing DWV Systems   85

Slow Drains   86
Bathtub Drains   88
Shower Drains   90
Toilet Drains   91
Frozen Drains   94
Overloaded Drains   95
Material Mistakes   96
The Sand Trap   96
The Building Drain That Wouldn't Hold Water   97
Duct Tape   98
Vent Problems   99

Chapter 5. Faucets   103

Kitchen Faucets   103
Lavatory Faucets   111
Bar Sinks and Laundry Tubs   112
Bathtubs and Showers   112
Wall-Mounted Faucets   114
Learn the Basics   115
Field Conditions   115

Chapter 6. Fixtures   119

What Fixtures Are Required?   119
Handicap Fixtures   123
Standard Fixture Installation Regulations   130
Troubleshooting Fixtures   145

Chapter 7. Water Heaters   157

Hot Water   158

Electric Water Heaters 163
Gas-Fired Water Heaters 167
Oil-Fired Water Heaters 170

Chapter 8. Well Systems 173

Shallow-Well Jet Pumps 174
Deep-Well Jet Pumps 177
Submersible Pumps 178
Pump Problems 181
Troubleshooting Jet Pumps 181

Chapter 9. Septic Systems 195

Simple Septic Systems 196
The Components 196
Chamber Systems 199
Trench Systems 200
Mound Systems 202
How Does a Septic System Work? 203
Septic Tank Maintenance 204
Piping Considerations 206
Gas Considerations 206
Sewage Pumps 207
Problems with a Leach Field 211

Chapter 10. Remodeling Reminders 215

Pipe Sizes 215
The Water Heater 217
Old Plumbing That Must Be Removed 217
Fixtures 218
Septic Systems 219
Rotted Walls and Floors 220
Cutting into Cast-Iron Stacks 221
Adding a Basement Bathroom 222
Size Limitations on Bathing Units 222
Cutoffs That Don't 222
Working with Old Pipes 223
Cast-Iron Pipe 227
Copper Water Pipes 229

Painted Copper Pipes   230

Chapter 11. Clogged Drains   233

Liquid and Powder Drain Openers   235
Water-Powered Drain Openers   236
Air-Powered Drain Openers   236
Closet Augers   237
Flat-Tape Snakes   237
Handheld Manual Spring Snakes   238
Big Drain Cleaners   238
Supersize Sewer Machines   239

Chapter 12. Materials Selection   241

What Is an Approved Material?   242
Water Service Pipe   244
Water Service Materials   244
Water Distribution Pipe   248
Water Distribution Materials   249
Drain-Waste-and-Vent Pipe   250
Storm Water Drain Materials   256
Subsoil Drains   260
Other Types of Materials   260
Connecting Materials   264

Chapter 13. Plumbing Appliances   269

Garbage Disposals   269
Pure Plumbing Problems with Disposals   273
Ice Makers   275
Dishwashers   278
Washing Machines   283

Chapter 14. Basics of Electricity   285

Color Codes   285
Wire Nuts   286
Box Selection   286
Common Heights and Distances   288
Wire Molding   289
Conduit   289

Snaking Wires   289
Putting Wire under the Screw   290
Ground Fault Interrupters   291
High-Voltage Circuits   291
Floor Outlets   291
Dishwashers   292
Garbage Disposals   292
Water Heaters   293

Chapter 15. Safety Procedures and Precautions   295

Dangers in Plumbing   295
General Safety   299
Tool Safety   302
Coworker Safety   312

Chapter 16. First Aid Basics   315

Open Wounds   316
Splinters and Such   319
Eye Injuries   320
Facial Injuries   321
Nosebleeds   321
Back Injuries   322
Leg and Foot Injuries   322
Hand Injuries   323
Shock   323
Burns   324
Heat-Related Problems   326

Chapter 17. Conversion Charts and Tables   329

Chapter 18. Sensible Pipe Sizing   361

Pipe Sizing   361
Sizing Building Drains and Sewers   365
Sizing Example   366
Horizontal Branches   367
Stack Sizing   368
Sizing Storm Water Drainage Systems   371
Sizing Vents   374

Vent Sizing Using Developed Length   375
Vents for Sumps and Sewer Pumps   381
Sizing Potable Water Systems   382
The Fixture-Unit Method   387

Chapter 19. Design Criteria   393

Grading Pipe   394
Supporting Pipe   394
Pipe Size Reduction   397
Underground Pipes   397
Fittings   397
Indirect Wastes   399
Special Wastes   402
Vent Installation Requirements   403
The Main Water Pipe   411
Pressure-Reducing Valves   412
Booster Pumps   413
Water Tanks   413
Water Conservation   414
Antiscald Precautions   415
Valve Regulations   415
Backflow Prevention   416
Hot Water Installations   419
Water Heaters   419
Common Sense   421

Index   423

Preface

Professional plumbers can make a lot of money. If you're a plumber, you should be making more money than the average worker. In fact, skilled plumbers can live lives that are extremely comfortable. If you are a master plumber and operate your own business, your income is unlimited.

Are you going to be the working plumber who puts in 40 hours a week for a fair paycheck and then hopes to retire when you're too weak to continue working? If you are a plumbing contractor, are you willing to make one-third of your potential earnings? Let me ask you a question. If you could improve your job performance and income by reading a book, would you do it? Most people would, and this book can be your ticket to bigger paychecks and a more profitable future.

I've been in the plumbing trade for more than 20 years. Starting as a helper making $2 an hour, I worked my way up the plumbing ladder quickly. When I was 21 years old, I was a job superintendent in charge of a 365-unit townhouse complex. A few years later, I opened my own business. It's been over 15 years since I struck out on my own, and my income continues to grow. Am I rich? It depends on your perspective; I'm not a millionaire, but I'm very comfortable. Plumbing has provided me and my family with a better-than-average life.

You are at an intersection in your career. Picking up this book is the first step in your journey to success. You can stop here, or you can move forward by reading the following pages and you can potentially reward yourself in ways beyond your most vivid imagination.

I would venture to say that over the last 15 years I probably have lost more money than you've earned. The losses were the result of inexperience and a lack of guidance. Fortunately, I learned from those mistakes, but they were very expensive lessons. By reading this book, you can gain the knowledge I have without losing hundreds of thousands of dollars. My personal losses have run in excess of $4 million. This is a staggering amount, but I'm still alive and well; if I can suffer such major losses and rebound with a successful present and promising future, you can do much more.

Let me give you some specifics about what this book can do for you. Do you ever feel confused about the installation of water distribution systems? Sooner or later, most plumbers do. The first chapter will give you plenty of advice on solving installation problems with water pipes.

After a potable water system is installed, there is always a chance that something will go wrong with it. Service plumbers have to deal with these problems all the time. Chapter 2 will show you how to troubleshoot and repair all potable water systems.

Water systems are only part of a plumbing job. Drains, wastes, and vents account for much plumbing system work. Chapters 3 and 4 delve into these areas and show you how to solve most DWV problems. All of your questions pertaining to DWV systems are answered in these chapters.

As you move along, Chaps. 5 and 6 cover faucets and fixtures. Water heaters are the subject of Chap. 7, and well systems Chap. 8. If you are working with wells, you will probably have to deal with septic systems, and you will find facts on these systems in Chap. 9.

Remodeling is a very profitable way of making money. Plumbers who work in remodeling generally do very well. Their work can be done in any weather and at any time of the year. Chapter 10 highlights the risks and rewards of remodeling.

Did you know that some plumbers make all their money by clearing drain stoppages? Well some do, and the money they can make might make you drool. There is major money in drain cleaning. Read Chap. 11 to see how cleaning drains can propel you into a higher tax bracket.

Material selection is important to a plumbing contractor. This is true for two reasons. One, the plumbing code requires certain materials to be used for specific types of jobs. Second, the materials you choose can affect the profit made on a job. Chapter 12 will show you what materials are appropriate for use in various circumstances and will detail the pros and cons of various types of materials.

Chapter 13 covers plumbing appliances, such as dishwashers and garbage disposals. You will learn to solve the subtle problems associated with plumbing applicances, problems that could cost you hours of time and hundreds of dollars a day.

Is a 12-gauge wire sufficient for a water heater? Do you need a disconnect box for an electric water heater? Are red wire nuts right for dishwasher connections? You will find answers to your electrical questions in Chap. 14.

Additional chapters in this book will guide you through safety issues, first-aid basics, conversion charts and tables, pipe sizing, and design criteria for plumbing systems. In short, if you need to know it, it's in this book.

Take a look at the table of contents. Turn a few pages of the book and sample the charts, tables, and illustrations. Read a paragraph or two. You'll see how reader-friendly the text is. This is a serious, down-to-earth bible for working plumbers. Without a doubt, your career will advance rapidly once you gain the wisdom of its contents.

Acknowledgments

I would like to acknowledge my parents, Maralou and Woody, for their positive influence on me over the last 40 years. They have provided invaluable support and assistance.

R. Dodge Woodson

1

Installing
Water Distribution
Systems

Installing water distribution systems is usually easier than installing a drain-waste-and-vent (DWV) system. For one thing, the pipes being installed with a water system are smaller. Also, the requirements for pipe grading are less stringent for water systems. Code requirements for water systems are easier for most plumbers to understand. When all elements are considered, a residential water system is pretty simple. Commercial jobs are more complex, but piping diagrams for commercial jobs make the installations easy if you are good at following directions.

Sizing water systems can be tricky. Most residential jobs are fairly simple, but sometimes a plumber can run into some complicated sizing situations. Since most residential water systems are designed by plumbers, not by engineers and architects, residential plumbers have to know how to interpret the plumbing code for sizing requirements.

Sizing is not the only part of installing a water system that can get tricky. It may be hard to decide what pipe will work best. The cost-effective routing of pipes in a layout is important to the plumber's profit picture. There's a lot to learn about water systems, so let's get started.

Potable Water

For what purposes do we require potable water? Potable water is needed for drinking, cooking, bathing, and the preparation of food and medicine. Potable water is also used for other activities, but those listed are the uses dealt with in the plumbing code. The plumbing of fixtures used for any of the above purposes must be done so that only potable water is accessible to them.

When must hot water be provided? Hot water is required in buildings where people work and in all permanent residences. What about cold water? Cold water must be supplied to every building that contains plumbing fixtures and is used for human occupancy.

How much water pressure is required? The water pressure for a water distribution system must be high enough to provide proper flow rates at each of the fixtures. Flow rates for various fixtures are determined by referring to the tables or text in your local code.

Generally, 40 pounds per square inch (psi) is considered adequate water pressure. If the incoming pressure is 80 psi or more, the pressure must be controlled with a pressure-reducing valve. This valve is installed between the water service and the main water distribution pipe. The device allows the water pressure to be kept at a lower value. There are, of course, exceptions to most rules; check your local code for the requirements in your jurisdiction.

On the subject of code requirements, note that three major plumbing codes are used in the United States. I have these codes broken down by location, and I refer to them as zones 1, 2, and 3 (Figs. 1.1, 1.2, and 1.3).

Water conservation

Water conservation is an issue that most codes address. The codes restrict the flow rates of certain fixtures, such as showers, sinks, and lavatories. An unmodified shower head can normally produce a flow of 5 gallons per minute (gpm). This flow rate is often reduced to 3 gpm with the insertion

Figure 1.1   States in zone 1.

Washington
Oregon
California
Nevada
Idaho
Montana
Wyoming
North Dakota
South Dakota
Minnesota
Iowa
Nebraska
Kansas
Utah
Arizona
Colorado
New Mexico
Indiana
Parts of Texas

Figure 1.2   States in zone 2.

Alabama
Arkansas
Louisiana
Tennessee
North Carolina
Mississippi
Georgia
Florida
South Carolina
Parts of Texas
Parts of Maryland
Parts of Delaware
Parts of Oklahoma
Parts of West Virginia

Figure 1.3   States in zone 3.

Virginia
Kentucky
Missouri
Illinois
Michigan
Ohio
Pennsylvania
New York
Connecticut
Massachusetts
Vermont
New Hampshire
Rhode Island
New Jersey
Parts of Delaware
Parts of West Virginia
Parts of Maine
Parts of Maryland
Parts of Oklahoma

of a water-saver device. The device is a small disk with holes in it.

Other water-saver regulations apply to buildings with public rest rooms. It is frequently required that all public-use lavatories be equipped with self-closing faucets. These faucets have restricted flow rates, and they will shut themselves off after use.

Urinals should have a flow rate of not more than $1\frac{1}{2}$ gallons (gal) per flush. Toilets generally must not use more than 4 gal of water during the flush cycle.

Antiscald devices

Antiscald devices are required on some plumbing fixtures. Different codes require these valves on different types of fixtures, and sometimes the maximum water temperature varies. Antiscald devices are valves or faucets that are specially equipped to prevent burns due to hot water. These devices come in different configurations, but they all have the same goal: to prevent scalding.

Whenever a gang shower is installed, such as in a school gym, antiscald shower valves should be installed. Some codes require antiscald valves on residential showers, but others don't. The maximum hot water temperature in some codes is 110°F. Other codes set it at 120°F.

Choosing the Type of Pipe and Fittings to Be Used

The three primary types of pipe used in above-grade water systems are polybutylene (PB), copper, and chlorinated polyvinyl chloride (CPVC). All these pipes provide adequate service as water distribution carriers. Which pipe you choose will be largely a matter of personal preference. Many homeowners opt for CPVC because they do not have to solder the joints. Soldering seems difficult to someone who has no experience with it. It seems easier to glue plastic pipe than to solder copper joints. Most people outside the trade never think of polybutylene; they simply don't know it exists.

Polybutylene has received a lot of publicity in recent months. Some people allege that the pipe is not dependable. Reports say that the pipe leaks at joints. One story I heard implied that chlorine in potable water deteriorated the pipe. Another story said that the problem was in the material used in making the fittings to connect the pipe. I'm not sure what the truth is.

I started using PB pipe a very long time ago. Back then, there were some problems with leaks when compression fittings were used, although I never had one. It has been more than 15 years since I started using PB pipe, and I've never had a single leak. I wish I could say the same about copper. It is my opinion that the problems associated with PB pipe are related to either the fittings being used or the plumbers making the connections. At the risk of going out on a shaky limb, I maintain that PB pipe is an excellent choice for water systems.

What pipe will you use? Before you answer this question, consider the differences in installation procedures for each type of pipe. The first consideration might be the cost of materials. Although plastic pipes are cheaper than copper pipes, plastic fittings are often more expensive than copper ones. At the end of plumbing an average home, there will not be a major difference in the material costs. The primary considerations are the durability of the pipe and the ease of installation.

CPVC Pipe

CPVC pipe has been used as potable water pipe for many years. The fittings for the pipe are installed with a solvent weld. You do not need soldering skills or equipment with CPVC. The pipe has some flexibility and is easy to snake through floor joists. It is suitable for hot and cold water applications and can be cut with a hacksaw. It is easy to install.

Professional plumbers I know, and have known, do not show great affection for CPVC. To make a waterproof joint, the pipe and fittings must be properly prepared. This preparation is very much like the procedures used with PVC drainage pipe. A cleaning solution must be applied to the pipe and fittings. Then a primer is applied to both the pipe and the fittings. Joints are made with a solvent or glue.

Going through all the steps of making the joint is a slow process. Since professionals seek to complete their jobs as quickly as possible, CPVC is not an ideal choice. Not only do you lose time with all the preparations, but also the joints cannot be disturbed for some time. If they are jarred before they have cured, the joints may suffer from voids that will leak. The cure time for the joint will vary depending upon the temperature.

The time lost waiting for the joints to set up is a drawback for professionals. When you work with polybutylene or copper, the time spent waiting to work with new joints is greatly reduced. This reason alone is enough to cause professionals to use a different pipe, but it is not the only reason professionals choose other pipes.

Even after a CPVC joint is made, it is not extremely strong. Any stress on the joint can cause it to break loose and leak. CPVC becomes brittle in cold weather. If the pipe is dropped on a hard surface, it can develop small cracks. These cracks often go unnoticed until the pipe is installed and tested. When the pipe is tested and leaks, it must be replaced. This means more lost time for the professional. Copper and polybutylene are not subject to these same cracks when dropped.

If the pipe or fitting has any water on it, the glue may not make a solid joint. With polybutylene, water on the pipe has no effect on the integrity of the joint. Water on the outside of copper pipe will turn to steam and normally does not cause a leak. The simple act of hanging the pipe is even a factor in making a choice. When you hang the pipe, you will probably use a hammer to drive the hanger into place. If the hammer slips and hits your pipe, CPVC may shatter. Copper may dent, but it will not shatter. Polybutylene will bounce right back into shape from a hammer blow. These may be small differences, but they add up to stack the deck against CPVC for professionals.

There is one advantage that CPVC has over copper for professional plumbing applications. It is not adversely affected by acidic water. When potable water is being provided by a well, acid in the water can cause copper pipe to leak. The acid causes pinhole leaks in copper after some time. Since CPVC is plastic, it is not affected by the acid.

Polybutylene Pipe

Polybutylene pipe is, in my opinion, a fine choice for water systems. There are people who disagree. As I said earlier, I've used PB pipe for over 15 years, and I've never had a problem with it.

If you are going to install PB pipe, learn how to do it correctly. This type of pipe is not like polyethylene. Special fittings and clamps should be used to ensure good joints. Polybutylene slides over a ridged insert fitting (I prefer copper fittings) and is held in place with a special clamp. These clamps are not adjustable, stainless-steel clamps, like those used on polyethylene pipe. The clamps used on polybutylene are solid metal and are installed with a special crimping tool. There are many advantages to working with polybutylene.

Polybutylene comes coiled in a roll. The pipe is extremely flexible and can be run very much as electric wire is. After holes are drilled, the pipe can be pulled through the holes in long lengths. The pipe is approved for hot and cold water and can be cut with a hacksaw, although special cutters do a neater job. In cold climates, polybutylene offers another advantage: It can expand a great deal before splitting during freezing conditions. Acidic water will not eat holes in PB pipe. It is difficult to think of a reason not to use polybutylene.

When houses are plumbed with polybutylene, the piping design is often different from that used when the house is plumbed with a rigid pipe. CPVC and copper installations incorporate the use of a main water pipe and several branches to feed individual fixtures. With polybutylene, most plumbers run individual lengths of pipe from each fixture to a common manifold. By installing the pipe in this way, there are no joints concealed in the walls or ceilings. This reduces the likelihood of a leak that will be difficult to access when the house is completed.

Basically, the water service or a main water pipe is run to the manifold location. The manifold receives its water from the water service or main and distributes it through the individual pipes. This type of installation requires more pipe than traditional installations, but polybutylene pipe is cheap. It is the fittings that are expensive, but in a manifold installation, very few fittings are used.

Manifold systems are convenient for service plumbers and homeowners. The fact that PB pipe can be installed with

very few concealed joints is an attractive attribute. PB pipe
may not always be the best pipe for a job, but it often is.

Copper Pipe

Copper has long been the leader in water distribution systems. While copper is expensive and has some drawbacks,
it remains the most common type of pipe in potable water
systems. Many homeowners specify copper in their building
plans, and old-school plumbers swear by it. As a seasoned
plumber, I like copper and use it frequently. However, I
seem to be using more and more PB pipe as people become
more accepting of it. Copper is a proved performer with a
solid track record.

When soldering joints on potable water pipes, you must
use a low-lead solder. In the old days, 50/50 solder was
used, but not today. Buy lead-free solder for your joints.
The old 50/50 solder is still available for making connections on nonpotable pipes, but don't use it on potable water
systems.

Copper is easy to cut when you use roller cutters (Fig.
1.4). To cut the pipe, simply place the pipe between the cutting wheel and the rollers, tighten the handle, and turn the
cutter. As you rotate the cutter around the pipe, tighten the
handle to maintain steady pressure. After a few turns, the
pipe will be cut. Be careful of jagged pieces of copper after
you cut the pipe. Be especially careful when you sand the
pipe to clean it. If the pipe does not cut evenly, there may
be sharp, jagged edges protruding from the edge of the
pipe. These copper shards can cut you or become embedded
in your skin.

Copper pipe is durable and is approved for hot and cold
water piping. Once you know how to solder, copper is easy
to work with. The pipe installs quickly and produces a
neat-looking job. Since the pipe is rigid, it can be installed
to allow the water system to drain. This is a factor in seasonal cottages and other circumstances where the water is
turned off for the winter.

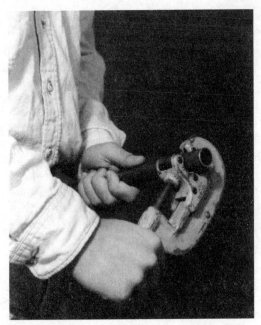

Figure 1.4 Roller cutters for cutting copper tubing and pipe.

The Pipe Decision

Which pipe to use is up to you. The average person would probably choose polybutylene. For a professional, it would be a toss-up between polybutylene and copper. The two factors that I would consider are the working conditions and type of plumbing being installed. Personally, I would not use CPVC. Consider all your options and make your choice based on those factors you feel strongly about.

Materials approved for water distribution

There are a number of materials approved for water distribution, but only a few are used frequently in new plumbing systems. Probably copper is still the number-one choice as a water distribution material, with PB pipe second and CPVC a distant third.

Materials approved for use in water distribution systems may vary slightly from code to code. The following information is valid in many jurisdictions, but remember to always check your local code before installing plumbing.

Galvanized steel pipe

Galvanized steel pipe is an approved water distribution pipe. It has been used for a very long time, and you will still find it in many older plumbing systems. It is not, however, used much in modern plumbing.

This pipe is heavy and requires threaded joints. The pipe tends to rust and deteriorate more quickly than other approved water distribution pipes. There are few, if any, occasions when galvanized steel pipe is used for modern plumbing installations.

Brass pipe

Brass pipe is approved for water distribution, but like galvanized steel pipe, it is rarely used.

Copper pipe

We talked about the use of copper as a water distribution pipe; it is, of course, approved for that application. There are, however, different types of copper. By types, I mean ratings of thickness. The two types most commonly used for water distribution are type L and type M. Type K copper is also fine for water distribution, but it is more expensive and usually the extra-thick pipe is not needed. Type M copper, the thinnest of the three, is not approved in all areas for water distribution. For years type M was the standard, but now type L is required by some administrative authorities.

Polybutylene pipe

Polybutylene is approved for water distribution, and we have already discussed its use. Will the recent accusations pertaining to PB pipe influence code officials to alter present code standards? Perhaps, but we will have to wait and see.

CPVC pipe

Chlorinated polyvinyl chloride, or CPVC as it is commonly called, is another approved material that we have discussed. You may find this pipe ideal, but I don't favor it. Try it and see what you think, but in experience I do better with other types of piping.

Sizing

The task of sizing water pipes strikes fear into the hearts of many plumbers. They take one look at the friction-loss charts in their code books and see some math formulas that read like a foreign language, and they run the other way. Sizing water pipes can be intimidating.

There are two approaches to sizing water pipe: the engineer's approach and the plumber's approach. Large jobs typically are required to be designed by an engineer or other suitable professionals, other than plumbers. Most plumbing codes will not allow plumbers to design complex water distribution systems. For most plumbers, myself included, this is a blessing.

Your local code book will provide information on how you can use formulas, friction charts, and other complicated aids to size water pipes; but it will also give you some easier ways around the problem.

Your code book will make clear what size pipe is needed for all common plumbing fixtures. Let me give you some idea of what you will find. One of the major codes allows fixture supplies with diameters of $\frac{3}{8}$ inch (in) to be used for the following fixtures: bidets, drinking fountains, single showers, lavatories, and water closets—with the exception of one-piece toilets. If you increase the fixture supply size to $\frac{1}{2}$ in, you can feed these fixtures: bathtubs, domestic dishwashers, hose bibs, residential kitchen sinks, laundry trays, service sinks, flush tank urinals, wall hydrants, and one-piece toilets.

By increasing the fixture size to $\frac{3}{4}$ in, you can connect to commercial kitchen sinks, flushing-rim sinks, and flush valve urinals. Stepping up to a 1-in pipe, you can serve toilets that use flush valves.

Now wasn't that easy? All you have to do is look at your code book to quickly see the minimum size requirements for your water fixture supplies. Up to the first 30 in of a section of pipe connecting to a fixture is considered a fixture supply. Once the supply is more than 30 in away from the fixture, it is considered a water distribution pipe. If you want to know the required flow rates and pressures for various fixtures, simply refer to your code book. All this information is available to you in an easy-to-understand format.

The next step is not quite as easy, but it still is not difficult. There are several methods for sizing water pipe. Most code books disclaim any responsibility for the examples they give on sizing. The examples are just that—examples, not carved-in-stone procedures. One of the easiest ways to size water pipe is with the use of the fixture-unit method, and that is the one we concentrate on. Let me show you how it works.

The fixture-unit method

The fixture-unit method of sizing water distribution pipes is pretty easy to understand. You will need to know the total number of fixture units that will be placed on your pipes. This information can be obtained from listings in your code book.

The total developed length of your water pipes must be known. If you are on the job, you can measure this distance with a tape measure. When you are working from blueprints, a scale rule will give you the numbers you need. The measurement begins at the location of the water meter.

You will also need to know the water pressure on the system. If you are not controlling the water pressure, as you might with a pressure-reducing valve or a well system, you can call the municipal water department to obtain the pressure rating on the water main.

To pinpoint the accuracy of your sizing, find out the difference in elevation between the water meter and the highest plumbing fixture. However, by being generous with your pipe sizes, you can get by without this information—most of the time.

The fixture units assigned to fixtures for the purpose of sizing water pipe will normally include both the hot and cold water demand of the fixture. The ratings for fixtures used by the public will be different from those fixtures installed for private use. All the information you need for this method of sizing is easily obtained from your local code book.

Now, let's size the water pipe for a small house. The house has one full bathroom and a kitchen. It also has a laundry hookup and one hose bib. We need to know what fixture-unit (FU) ratings to assign to these fixtures. A quick look at a code book might give us the following ratings:

*Bathtub: 2 FU   *Lavatory: 1 FU

*Toilet: 3 FU   *Sink: 2 FU

*Hose bib: 3 FU   *Laundry hookup: 2 FU

Okay, now we know the total number of fixture units. The working pressure on this water system is 50 psi. The total developed length of piping, from the water meter to the most remote fixture, is 95 feet (ft).

We now have all we really need to size water pipe, with a little help from a table in our code book. When we look for the sizing table in the code book, we see different tables for different pressure ratings. We choose the table that matches our rating of 50 psi.

There will be columns of numbers to identify fixture units and developed lengths of pipe. We will see sizes for building water supplies and branches. Even the sizes for water meters will be available. All we have to do is put our plan into action.

Our water meter for this example is a $\frac{3}{4}$-in meter. First we have to size the water service. How many fixture units will be on the water service? When we count all the fixture units, the total is 13. The pipe has to run 95 ft, so what size does it have to be? Looking at the sizing table in your code book, you will quickly see that the required size is $\frac{3}{4}$-in water service. I was able to determine this just by running my finger down and across the sizing table.

The sizing table is based on the length of the pipe run, the water pressure, and the size of the water meter. Knowing those three variables, you can size pipe quickly and easily. The same cross-referencing that was done to size the water service also works for sizing water distribution pipes.

For example, assume that the most remote fixture in the house is 40 ft from the end of the water service. When I look at my table, I see that a $\frac{3}{4}$-in pipe is required to begin the water distribution system. As the main $\frac{3}{4}$-in pipe reaches fixtures and lowers the number of fixture units remaining in the run, the pipe size can be reduced. As a rule of thumb, when there are only two fixtures remaining to be served on a residential water main, the pipe size can be reduced to a $\frac{1}{2}$-in pipe.

Let's say the $\frac{3}{4}$-in pipe is running down the center of the home. The kitchen sink is perpendicular to the main water line and lies 22 ft away. The sink has a fixture-unit demand of 2. Checking my sizing table, I see that the pipe branching off the main to serve the sink can be a $\frac{1}{2}$-in pipe.

Once you sit down with your code book and work with the fixture-unit method of pipe sizing, you will see how easy it is. Now, we are done with our sizing exercise, and we are going to move on to some techniques used to install water distribution pipes.

Installing the Potable Water System

Regardless of how the connections are made, CPVC and copper systems are installed according to the same principles. Polybutylene systems could be installed along the same lines as copper, but a manifold installation makes more sense with polybutylene. PB pipe is available in semi-rigid lengths, in addition to coils.

Is a water service part of a water distribution system? It is part of the water system, but it is not normally considered to be a part of a water distribution system. This can be confusing, so let me explain.

A water service does convey potable water to a building, but it is not considered a water distribution pipe. Water services fall into a category all their own. Water distribution pipes are the pipes found inside a building. In water services the majority of the pipe length is outside a building's foundation. They are usually buried in the ground. Water distribution pipes are installed within the foundation of the building and don't normally run underground or outside a foundation. However, they can, as in the case of a one-story building where the water distribution pipes are buried under a concrete floor. In any event, water services and water distribution pipes are dealt with differently by the plumbing codes.

Why are water distribution pipes considered different from water services? Water distribution pipes are dealt with differently because they serve different needs. Water distribution pipes pick up where water services leave off. The pipes distributing water to plumbing fixtures do not have to meet the same standards as underground water services.

For one thing, the water pressure on a water distribution pipe is often less than that of a water service. Most homes have water pressure ranges between 40 and 60 psi. A water service from a city main could easily have an internal pressure of 80 psi, or more.

Water services are also buried in the ground. Water distribution pipes rarely are. Underground installation can affect the types of pipes that are approved for use. Another big difference between water service pipes and water distribution pipes is the temperature of the water they contain. Water services deliver cold water to a building. Water distribution pipes normally distribute both hot and cold water. Since most codes require the same type of pipe to be used for the cold water lines as for hot water pipes, the temperature ranges can limit the applications of some types of materials. This is a significant factor when you are choosing a water distribution material.

Valves

Many valves are required in a legal water distribution system. The types of valves used and the places where they are required are mandated by the code. For example, a full-open valve is required near the origination of a water service. In the case of a city water hookup, the valve is supposed to be installed near the curb of the street. Water services extending from a well are not required to have such a valve at their origination. A full-open valve is a valve like a gate valve or ball valve. When these valves are open, water has the full diameter of the valve to flow through, rather than a restricted opening, as in valves with washers and seats.

A full-open valve is required on the discharge side of every water meter. The main water distribution pipe must be fitted with a full-open valve where it meets the water service. In nonresidential buildings, a full-open valve must be installed at the origination of every riser pipe. In these nonresidential buildings, a full-open valve is required at the top of all drop-feed pipes.

If a water supply pipe is feeding a water tank, such as a well pressure tank, a full-open valve is required. All water supplies to water heaters must be equipped with full-open valves. If a building contains multiple dwellings, the water supplies for each dwelling must be equipped with full-open valves. Remember that not all codes have the same requirements. Check your local code for applicable restrictions.

The valves we are about to discuss are not required to be full-open valves. The valves used in the following locations are usually either stop-and-waste valves or supply-stop valves. Valves are required on all supply pipes feeding sill cocks and hose bibs. If a water supply is serving an appliance, such as an ice maker or dishwasher, a valve is required.

Some types of buildings are required to have individual cutoffs on the supplies to all plumbing fixtures. Other types of buildings are not. It is generally standard procedure to install cutoffs on all water supplies, with the possible exception of the pipes feeding tub and shower valves.

When a valve is required by the code, that valve must be accessible. It may be concealed by a door that opens or some other means of accessible concealment, but it must be accessible. Stop-and-waste valves are normally not approved for use below ground.

Backflow Protection

Backflow protection for water systems has become a major issue in the plumbing trade. The forms of protection required involve such issues as cross-connections, backflow, and vacuums. If backflow preventers are not installed, entire water mains and water supplies can be at risk of contamination.

Many types of cross-connections are possible. For example, hot water may pass into cold water pipes through a cross-connection. This shouldn't contaminate the system, but it can create a health risk. Someone who turns on what is believed to be cold water and discovers that it's hot water could be burned. Normally, this type of cross-connection is more of a nuisance than a hazard, but it does have the potential to cause injury.

A much more serious type of cross-connection might occur in a commercial photography processing plant. What would happen if the contents of a pipe conveying photography chemicals were allowed to mix with the potable water system? The results could be quite serious, indeed.

Cross-connections are normally created by accident, but sometimes they are designed. If a cross-connection of some sort is to be installed, it must be protected with the proper devices to avoid unwanted mixing. Few, if any, codes will allow a cross-connection between a private water source, such as a well, and a municipal water supply. Check your local code requirements for backflow prevention devices.

An air gap

An *air gap* is the open airspace between the flood-level rim of a fixture and the bottom of a potable water outlet. For example, usually there is an air gap between the outlet of a spout on a bathtub and the flood-level rim of the tub.

By having such an air gap, nonpotable water, such as dirty bath water, cannot be sucked back into the plumbing system through the potable water outlet. If the tub spout were mounted so that it was submerged in the tub water, the dirty water could be pulled into the water pipes if a vacuum formed in the plumbing system.

The amount of air gap required on various fixtures is determined by the plumbing code. For example, in a table or in the text of your code, you might find that the tub spout opening must not be closer than 2 in to the flood-level rim. The same rule for lavatories might require the spout of the faucet to remain at least 1 in above the flood-level rim.

There is another type of air gap used in plumbing. This fitting accepts the waste from a dishwasher for conveyance to the sanitary drainage system. The device allows the water draining out of the dishwasher to pass through open airspace on its way to the drainage system. Since the water passes through open air, it is impossible for wastewater from the drainage system to be siphoned back into the dishwasher.

Backflow preventers

Backflow preventers are devices that prohibit the contents of a pipe from flowing in a direction opposite its intended flow. These protection devices come in many shapes and sizes, and they cost from a few dollars to thousands of dollars. Backflow preventers must be installed in accessible locations.

When a potable water pipe is connected to a boiler to provide a supply of water, the potable water pipe must be equipped with a backflow preventer. If the boiler has chemicals in it, such as antifreeze, the connection must be made through a backflow preventer of an air gap that works on the principle of reduced pressure.

Anytime a connection is subject to back pressure, the connection must be protected with a reduced-pressure type of backflow preventer.

Check valves

Check valves have an integral section, something of a flap, that allows liquids to flow freely in one direction but not at all in the opposite direction. Check valves are required on potable water supply pipes that are connected to automatic fire sprinkler systems and standpipes.

Vacuum breakers

Vacuum breakers are installed to prevent a vacuum from being formed in a plumbing system that could result in contamination. Some of these devices mount in line on water distribution pipes, and others screw onto the threads of sill cocks and laundry tray faucets.

When vacuum breakers are required, the critical level of the device is usually required to be set at least 6 in above the flood-level rim of the fixture. What is the critical level?

The *critical level* is the point at which the vacuum breaker would be submerged, prior to backflow.

Normally, any type of fixture that is equipped with threads for a garden hose must be fitted with a vacuum breaker. Some sill cocks are available with integral vacuum breakers. For the ones that are not, small vacuum breakers that screw onto the hose threads are available. After the devices are secured on the threads, a setscrew is turned into the side of the device. The screw is tightened until its head breaks off. This ensures that the safety device cannot be removed easily.

Since vacuum breakers have an opening that leads to the potable water system, they cannot be installed where toxic fumes or vapors may enter the plumbing system through the openings. One example of a place where they could not be installed is under a range hood.

The starting point for your potable water system will be where the water service enters the house (see Figs. 1.5 and 1.6). Connect your main cold water pipe to the water service. In most places you will be required to install a backflow preventer on the water service pipe. If your water service is polyethylene pipe, it must be converted to a pipe rated for hot and cold water applications. Polyethylene cannot handle the high temperature of hot water. The code requires you to use the same type of pipe for both hot and cold water. Since polyethylene cannot be used for hot water, it cannot be used as an interior water distribution pipe in a house that has hot water.

You can convert the polyethylene pipe to the pipe of your choice with an insert adapter. The insert portion fits inside the polyethylene pipe and is held in place by stainless-steel clamps. The other end of the fitting is capable of accepting the type of pipe you are using for water distribution. In most cases the adapter is threaded on the conversion end. The threads allow you to mate any type of female adapter to the insert-by-male adapter.

Once the conversion is made, install a gate valve on the water distribution pipe. When required, install a backflow preventer following the gate valve. It is a good idea to

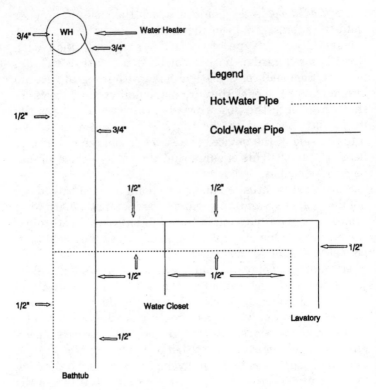

Figure 1.5 Water pipe riser diagram.

install another gate valve after the backflow preventer. By isolating the backflow preventer with the two gate valves, you can cut off the water on both sides of the backflow preventer. This option will be appreciated if you must repair or replace the backflow preventer.

As you bring pipe out of the second gate valve, consider installing a sill cock near the rising water main. If the location is suitable, you will use a minimum amount of pipe for the sill cock. The main water distribution pipe for most houses will be a ¾-in pipe. If you have already designed your water distribution layout, install the pipe accordingly. If you haven't made a design, do it now.

The ¾-in pipe should go to the inlet of the water heater,

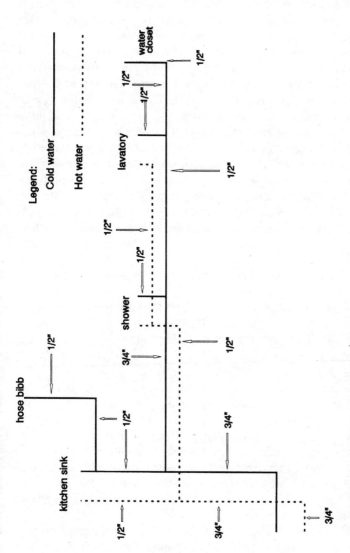

Figure 1.6 Diagram of efficient water supply.

undiminished in size. You may choose to branch off the main to supply fixtures with water as the main goes to the water heater. This is acceptable and economical.

Water pipe will normally be installed in the floor or ceiling joists so that they can be hidden from plain view. When possible, keep the pipe at least 2 in from the top or the bottom of the joists. Keep in mind that your local building code will prohibit you from drilling, cutting, or notching joists close to their edges. Keeping pipes near the middle of studs and joists reduces the risk of the pipes being punctured by nails or screws. If the pipe passes through a stud or joist where it might be punctured, protect it with a nail plate.

Proper pipe support is always an issue that all plumbers must be aware of. Copper tubing with a diameter of $1\frac{1}{4}$ in or less should be supported at intervals not to exceed 6 ft. Larger copper can run 10 ft between supports. These recommendations are for pipes installed horizontally. If the pipes are run vertically, both sizes should be supported at intervals of no more than 10 ft.

If your material of choice is CPVC, it should be supported every 3 ft, whether installed vertically or horizontally. Polybutylene pipe that is installed vertically must be supported at intervals of 4 ft. When installed horizontally, PB pipe should have a support every 32 in.

Avoid placing water pipes in outside walls and areas that will not be heated, such as garages and attics. The main delivery pipe for hot water will originate at the water heater. When you are piping the water heater, the cold water pipe should be equipped with a gate valve before it enters the water heater. Many places require the installation of a vacuum breaker at the water heater. These devices prohibit backsiphoning of the contents of the water heater into the cold water pipes.

Do not put a cutoff valve on the hot water pipe leaving the water heater. If a valve is installed on the hot water side, it could become a safety hazard. Should the valve become closed, the water heater could build excessive pressure. The combination of a closed valve and a faulty relief valve could cause a water heater to explode.

During the rough-in, you will not normally be installing valves on various fixtures. Most fixtures get their valves during the finish plumbing stage. If you install your sill cocks or hose bibs during the rough-in, install stop-and-waste valves in the pipe supplying water to the sill cock or hose bib. Stop-and-waste valves have an arrow on the side of the valve body. Install the valve so that the arrow is pointing in the direction in which the water flows to the fixture.

Rough-in numbers

Refer to a rough-in book (available from plumbing suppliers for various brands of fixtures) to establish the proper location for your water pipes to serve the fixtures. The examples given here are common rough-in numbers, but your fixtures may require a different rough-in. The water supplies for sinks and lavatories are usually placed 21 in above the subfloor. Kitchen water pipes are normally set 8 in apart, with the centerpoint being the center of the sink. Lavatory supplies are set 4 in apart, with the center being the center of the lavatory bowl.

Most toilet supplies will rough in at 6 in above the subfloor and 6 in to the left of the closet flange, as you face the back wall. Shower heads are routinely placed 6 ft 6 in above the subfloor. They should be centered above the bathing unit. Tub spouts are centered over the tub's drain and roughed in 4 in above the tub's flood-level rim. The faucet for a bathtub is commonly set 12 in above the tub and centered on the tub's drain. Shower faucets are generally set 4 ft above the subfloor and centered in the shower wall. If you ask your supplier for a rough-in book, the supplier should be able to give you exact information for roughing in your fixtures.

Secure all pipes near the rough-in for fixtures. When you rough in for a shower head, use a wing ell. These ells have 0.5-in female threads to accept the threads of a shower arm. The wing ell has an ear on each side that allows you to screw the ell to a piece of backing in the wall. Securing the ell is convenient for later installation of the shower arm. Where necessary, install backing in the walls to which all water pipes can be attached. It is very important to have all pipes firmly secured.

If you are soldering joints, remember to open all valves and faucets before soldering them. Failure to do this could damage the washers and internal parts once heat is applied. Consider installing air chambers to reduce the risk of noisy pipes later. The air chamber can be made from the same pipe as you use in the rest of the house. Air chambers should be at least one pipe size larger than the pipes they serve. A standard air chamber will be about 12 in tall (Fig. 1.7).

Installing cutoff valves in the pipes feeding the bathtub

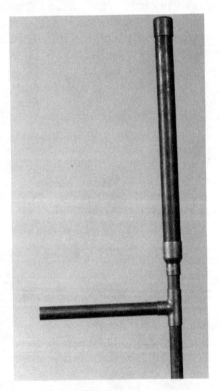

Figure 1.7 Air chamber.

or shower is optional. Most codes do not require the installation of these valves, but the valves come in handy when
you have to work on the tub or shower faucets in later
years. Stop-and-waste valves are typically used for this
application. Be sure to install the valves where you will
have easy access to them after the house is finished.

Avoid extremely long, straight runs of water pipe. If you
are running the pipe a long way, install offsets in it. The
offsets reduce the risk of annoying water hammer later. A
little thought during the rough-in can save you a lot of
trouble in the future.

A Manifold System

The rules for plumbing a water distribution system are
basically the same for all approved materials. But when
polybutylene is used, often it makes more sense to use a
manifold than to run a main with many branches. One of
the most desirable effects of this type of installation is the
lack of concealed joints. All the concealed piping is run in
solid lengths, without fittings.

To plumb a manifold system, run the water main to the
manifold. The manifold can be purchased from a plumbing
supplier. It will come with cutoff valves for the hot and cold
water pipes. The cold water comes into one end of the manifold and goes out the other to the water heater. The hot water
comes into the manifold from the water heater. The manifold
is divided into hot and cold sections. The pipe from each fixture connects to the manifold. When there is a demand for
water at a fixture, it is satisfied with the water from the manifold. Once you have both the hot and cold supply to the manifold, all you have to do is run individual pipes to each fixture.

You can snake coiled polybutylene pipe through a house
in much the same way as you would electric wires. By
installing the water distribution system in this way, you
save many hours of labor. You also eliminate concealed
joints, and you can cut off any fixture with the valves on

the manifold. This type of system is efficient, economical, and the way of the future.

Tips for Underground Water Pipes

To close this chapter, let me give you a few tips for underground water pipes. When you are ready to run water pipe below grade, there are a few rules you must obey. Pipe used for the water service must have a minimum working pressure of 160 psi at 73.4°F. If the water pressure in your area exceeds 160 psi, the pipe must be rated to handle the highest pressure available to the pipe. If the water service is a plastic pipe, it must terminate within 5 ft of its point of entry into the home. If the water service pipe is run in the same trench as the sewer, special installation procedures are required (refer to your local plumbing code).

When running the water service and the sewer in a common trench, you must keep the bottom of the water service at least 12 in above the top of the sewer. The water service must be placed on a solid shelf of firm material to one side of the trench. The sewer and the water service should be separated by undisturbed or compacted earth. The water service must be buried at a proper depth to protect it from freezing temperatures. The depth required will vary from state to state. Check with the code enforcement office for the proper depth in your area.

Copper pipe run under concrete should be type L or type K copper. Where copper pipe will come into direct contact with concrete, it should be sleeved to protect the copper. Concrete can cause a damaging reaction when in direct contact with copper. The pipe may also vibrate during use and wear a hole in itself by rubbing against rough concrete. All water pipe should be installed so as to prevent abrasive surfaces from coming into contact with the pipe.

Rules for underground piping are different from those for aboveground systems. Consult your local plumbing code to identify differences which affect plumbers in your area.

2

Troubleshooting and Repairing Water Distribution Systems

Troubleshooting water distribution systems can be very simple. Plumbers can often look for the source of a leak and find it by following the sound of rushing water or a trail of water stains. If finding the problem were always this easy, you wouldn't need to read this chapter. However, troubleshooting a water system can require a lot of time and thought. Some problems disguise themselves well and are difficult to find.

Plumbers who have developed strong troubleshooting skills can usually guess what's wrong on a job before they ever arrive at the site. However, this type of knowledge doesn't come quickly or easily. To become a proficient troubleshooter, you must learn to use the principles of a methodical approach in problem-solving situations. Experience is one of the best teachers, but this chapter can help you jump-start your learning experience.

What's the first type of problem that you think of when you consider trouble with a water system? Is it water leaks? Problems with leaks are common. Fortunately, the sources of leaks in a water system are often easy to locate. And correcting such a problem is not usually difficult. But how about water pressure? Many homeowners complain about their water pressure, usually because it is not adequate to satisfy them. What would you look for if a customer complained of low water pressure? There are many possible causes. Let's look at some of them.

Low Water Pressure

Low water pressure often becomes a problem in plumbing systems. Trying to take a shower in a home with low water pressure is very frustrating, and waiting several minutes for a water closet to refill its tank can be nerve-racking. There are many reasons why a lack of water pressure will induce customers to call in professional plumbers.

What causes low water pressure? Many factors can cause the water pressure level to be less than desirable. Buildings served by water pumps can be plagued by low water pressure when there is a problem with a pressure tank or the pressure switch. Homes that are connected to city water supplies may have pressure-reducing valves that need to be adjusted. Properties with old galvanized steel piping could be suffering from rust obstructions in the pipes.

Pressure tanks

Properties that obtain their water from private water sources depend on pressure tanks to give them the working pressure they want from their plumbing systems. If the pressure tanks become waterlogged, they cannot produce the type of water pressure they are designed to provide.

A waterlogged pressure tank has a symptom that is hard for a serious troubleshooter to miss. When a pressure tank is waterlogged, the well pump will cut in very frequently, often every time a faucet is opened. These tanks sometimes provide adequate pressure, even though they are forcing the pump to work much harder than it is intended to. At other times, a loss in pressure will be noticeable.

If you are faced with a building that has unsatisfactory water pressure and a well pump that cuts in frequently, take a close look at the pressure tank. When a tank is suspected of being waterlogged, it should be drained, recharged with air pressure, and then refilled with water. This is a simple process, and it can solve your pressure problems.

Pressure switches

Properties that are served by private wells depend on both pressure tanks and pressure switches. If the cut-in pressure on a well system is not set properly, the pressure may drop to unacceptable levels before the pump will produce more water.

A fast visual inspection of the pressure gauge on the well system will tell you whether the system is maintaining a satisfactory working pressure. It is important, however, to make sure there is demand from a plumbing fixture while you are observing the pressure gauge. If no plumbing fixtures are being called upon for water, the pressure gauge will remain static at its highest level. To be sure the system is maintaining an acceptable working pressure, you must put a demand on the system.

A pressure gauge that shows a sharp fall in pressure before the pump cuts in indicates that the cut-in pressure is set too low. Adjusting the spring-loaded nut in the box of the pressure switch will correct this problem.

Pressure-reducing valves

Pressure-reducing valves that are not set properly can cause low water pressure. If these valves are not adjusted to the proper settings, they can reduce the waterflow to little more than a trickle.

Sometimes the adjustment screws on pressure-reducing valves are turned down too tightly. In a low-pressure situation where a pressure-reducing valve is involved, check the level of the screw setting. Turning the adjustment screw counterclockwise will increase the water pressure on the system.

Pressure-reducing valves are not often the cause of a pressure problem, but they can be. Perhaps the adjustment screw is set improperly, or maybe the entire valve is bad. If you cut the water off at the street connection, you can check the building pressure by removing the pressure-reducing valve and installing a pressure gauge on the piping. This will eliminate, or confirm, any doubts about the pressure-reducing valve.

Galvanized pipe

If you have been in the plumbing trade long enough to have worked with much galvanized pipe, you know how badly it can rust, corrode, and build up obstructions. Anytime you are working with a water distribution system that is made up of galvanized pipe, do not be surprised to find low water pressure.

I have cut out sections of galvanized pipe where the open pipe diameter wasn't large enough to allow the insertion of a common drinking straw. When the open diameter of the pipe is constricted, the water pressure must drop. Any long-time plumber will tell you that galvanized water pipe is a nightmare waiting to happen.

Most plumbing systems plumbed with galvanized pipe are equipped with unions on various sections of the piping. If you cut off the water and loosen the unions, you will likely find the cause of the low pressure. A quick look at the interior of the pipe will probably be all that is needed to convince you.

The only real solution to low pressure due to blocked galvanized pipe is the replacement of the water piping. This can be a big and expensive job, but it is the only way to solve the problem positively.

Undersized piping

Undersized piping can create problems with water pressure. If the pipes delivering water to fixtures are too small, the fixtures will not receive an adequate volume of water at a desired pressure.

Hard water

Hard water can be a cause of reduced water pressure. The scale that builds up in hard water on the inside of water pipes, fixtures, and tanks can reduce water pressure by a noticeable amount. When a job is giving trouble with low pressure and there is no sign of a cause, inspect the interior of some piping to see if it is being blocked by scale buildup.

Clogged filters

Clogged in-line filters are notorious for their ability to restrict water flow. People have these little filters put in to trap sediment, and they do. They trap the sediment so well that they eventually clog up and block the normal flow of water, thereby reducing water pressure.

When you respond to a low-pressure call, ask the property owner if any filters are installed on the water distribution system. If so, check the filter to see whether it needs to be replaced.

Stopped-up aerators

One of the most common—and simplest to fix—causes of low water pressure is stopped-up aerators. The screens in the aerators stop up frequently when a house is served by a private water supply. Iron particles and mineral deposits are usually the cause of stoppages.

Aerators can be removed and sometimes can be cleaned, but many times they must be replaced. As long as you have an assortment of aerators on your service truck, this is one problem that is both easy to find and easy to fix.

Excessive Water Pressure

Excessive water pressure in a water distribution system can be dangerous and destructive. When fixtures are under too much pressure, they do not perform well over extended periods. The O-rings, stems, and other components of the plumbing system are not usually meant to work with pressures exceeding 80 pounds per square inch (psi).

Extreme water pressure can be dangerous to users of the plumbing system. For example, I once worked for a homeowner who had cut herself badly at the kitchen sink as a result of high water pressure. She was holding a drinking glass under the kitchen faucet when she turned on the water to fill the glass. The water rushed out with such force that it knocked the glass out of her hand. Glass shattered on impact with the sink, and pieces of the broken glass sliced into the woman's hand and arm.

High water pressure is easy to control. The pressure can be reduced with a pressure-reducing valve and, in the case of well systems, with adjustments to the pressure switch.

Troubleshooting high water pressure is simple. Just look at the pressure gauge on the well system, or install a pressure gauge on a faucet, usually a hose bib. If the pressure is higher than 60 psi, normally it should be lowered. Most residential properties operate well with a water pressure of between 40 and 50 psi.

If you're working with a well system, adjust the cutout setting in the pressure switch. In the case of city water service, you will have to install a pressure-reducing valve. If a pressure-reducing valve is already in place, you can lower the system pressure by turning the adjustment screw clockwise.

Water Hammer

Noisy pipes, resulting from water hammer, can make living around a plumbing system very annoying. Sometimes the pipes bang, squeak, and chatter. This condition can be so severe that living with the noise is almost unbearable. What causes noisy pipes? Water hammer is one reason pipes play their unfavorable tunes, but it is not the only reason.

Not all pipe noises are the same. The type of noise you hear is a strong indication of the type of action needed to solve the problem. You must listen closely to the sounds in order to diagnose the problem properly.

What will you be listening for? The tone and type of noise being made will be all that you may have to go on, in solving the problem of water hammer. Let's look at various noises individually and see what they mean and how you can correct the problem.

Water hammer is the most common cause of noisy pipes in a water distribution system. Banging pipes are a sure sign of water hammer. Can water hammer be stopped? Yes, there are ways to eliminate the actions and effects of water hammer, but implementing the procedure is not always easy.

What causes water hammer? Water hammer occurs most often with quick-closing valves, such as ball cocks and washing machine valves, but it can be a problem with other fixtures. The condition can worsen when the water distribution pipes are installed with long, straight runs. When the water is shut off quickly, it bangs into the fittings at the end of the pipe run or at the fixture and produces the hammering or banging noise. The shock wave can produce some loud noises.

If a plumbing system is under higher-than-average pressure, it is more likely to suffer from water hammer. There are several ways to approach the problem to eliminate the banging. To illustrate these options, I would like to put the problems into the form of real-world situations. Allow me to give you an example from one of my company's past service calls.

A toilet

This first example of a water hammer problem involves a troublesome toilet. The residents of the home where the toilet was located hated the banging noise their water pipes made on most occasions when the toilet was flushed. The toilet was located in the second-floor bathroom, and when it was flushed, it would rattle the pipes all through the home. One day the homeowners decided they could not live with the problem any longer. They were fed up with the annoying banging every time the toilet was flushed.

The couple called in a plumber to troubleshoot the problem. The plumber looked over the situation and went about his work, trying to locate the cause of the problem. He knew they were suffering from a water hammer situation, but he wasn't sure how to handle the call. In fact, the plumber gave up, washed his hands of the job, and left.

Distraught, the homeowners called my company. One of my plumbers responded to the call and quickly assessed that the upstairs toilet was the culprit. He recommended installing an air chamber in the wall, near the toilet.

At first, the couple was reluctant to give permission to cut into the bathroom wall. Under the circumstances, however, there were not many viable options, so the homeowners consented.

The plumber made a modest cut in the bathroom wall, around and above the closet supply, and installed an air chamber. This particular homeowner wanted to know every move that was being made, and he was not too sure the plumber was doing anything that would really improve the situation.

After the air chamber was installed, the plumber activated the system and began to flush the toilet. After several sequences of flushing the pipes remained quiet; there was no banging. The homeowner had been doubtful of the plumber's decision, but he did admit the work solved the problem, and he was very happy.

Air chambers are frequently all that is needed to reduce or eliminate banging pipes. The plumber was correct in his diagnosis of the problem and in effecting an efficient cure for the problem. It is possible, however, that the problem could have been solved by installing a master unit for a water hammer arrestor in the basement of the home, eliminating the need to cut into the bathroom wall. But there is no guarantee that a master unit would have controlled the situation on the second floor.

The plumber made a wise and prudent decision. By installing the air chamber at the fixture, satisfaction was practically guaranteed. While the master unit may have worked, there was some risk that it wouldn't. If the homeowner had objected to opening the bathroom wall, the plumber would have installed a master unit in an attempt to solve the problem.

Existing air chambers

Sometimes existing air chambers become waterlogged and fail to function properly. This problem is not uncommon, but it is a bother. If a plumbing system equipped with air chambers is being affected by water hammer, you must recharge the existing air chambers with a fresh supply of air. This is a simple process.

To recharge air chambers, first drain the water pipes of most of their reserve water. This can be done by opening a faucet or valve that is at the low end of the system and one that is at the upper end of the system.

Once the water has drained out of the water distribution pipes, close the faucets or valves and turn the water supply back on. As the new water fills the system, air will be replenished in the air chambers. As easy as this procedure is, it is all that is required to recharge waterlogged air chambers.

Unsecured pipes

Unsecured pipes are prime targets when you are having a noise problem. If the pipes that make up the water distribution system are not secured properly, they are likely to vibrate and make all sorts of noise. This not only is annoying to the ears, but also can be damaging to the pipes.

Pipes that are not secure in their hangers can vibrate so much that they wear holes in themselves. If enough stress is present, the connection joints may even be broken. Local plumbing codes dictate how far apart the hangers for pipe must be. If the hangers fall within the guidelines of the local code and secure the pipes tightly, no problems should exist. However, when the hangers are farther apart than they should be, are the wrong size, or are not attached firmly, trouble can develop.

Pipes that are not secured in the manner described by the plumbing code may create loud, banging sounds. This noise imitates the noise made by a system that is experiencing water hammer. Squeaking and chattering noises may also be present when pipes are not secured properly.

The problem of poorly secured pipes is easy to fix, if you have access to the pipes. Unfortunately, pipes and their hangers are often concealed by finished walls. Sometimes the problem pipes can be accessed in basements or crawl spaces, but often they are inaccessible, unless walls and ceilings are destroyed.

Once you have access to the pipes that are not properly secured, add hangers or tighten the existing hangers. This is fine if you can get to the pipes easily, but it is a hard sell to a homeowner who doesn't want the walls and ceilings of a home cut open. There is, however, no other way to solve the problem.

That squeaking noise

That squeaking noise you often hear from water pipes is almost always a hot water pipe. Whether the water distribution pipe is made of copper or plastic, it tends to expand when it gets hot. As hot water moves through the pipe, the pipe expands and creates friction against the pipe hanger.

The expansion aspect of hot water pipes is not practical to eliminate, but you can do something about the squeaking noise. The simple solution, when you have access to the pipe and hanger, is to install an insulator between the pipe and the support. This can be a piece of foam, a piece of rubber, or any other suitable insulator. Your only goal is to prevent the pipe from rubbing against the hanger when the pipe expands.

Chattering

The chattering heard in plumbing pipes is not due to cold temperatures; it is due to problems in the faucets or fixtures. Although this is more of a faucet and valve problem than a water distribution problem, we cover it here, since the noise is often associated with water pipes.

When you hear a chattering sound in a plumbing system, inspect the faucet stems and washers at nearby faucets. You should find that the faucet washer has become loose and is being vibrated, or *fluttered,* as some plumbers say, by the water pressure between it and the faucet seat.

To solve the problem of chattering pipes, tighten the washers in the faucets. Once the washers are tight, the noise should disappear.

Muffling the system at the water service

When you have a plumbing system that is banging due to water hammer, you can try muffling the system at the water service. This procedure doesn't always work, but sometimes it does.

Access to piping in many buildings is not readily available. Under these conditions the only easy place to eliminate water hammer is at the water service or main water distribution pipe. By installing air chambers or water hammer arrestors on sections of the available pipe, you may be able to eliminate the symptoms of water hammer throughout the building.

Installing an air chamber or water arrestor on the water service or main water distribution pipe is not a big job. Once the water is cut off, simply cut in some tee fittings and install the air chambers or arrestors. The risers that accept these devices should be at least 2 ft high, when possible.

Depending upon the severity of the water hammer problem, several devices may have to be installed at different locations along the water piping. Typically, one device at the beginning of the piping, one device at the end of the piping, and devices installed at the ends of branches will greatly reduce, if not eliminate, the problems associated with water hammer.

Adding offsets

Adding offsets to long runs of straight piping is another way to reduce the effects of water hammer. Long runs of piping invite the slamming and banging noise of water hammer. If you cut out the straight sections and rework them with some offsets, you increase the odds of eliminating the banging noises.

Leaks in Copper Piping

Copper leaks are the most common type found in the piping of water distribution systems. Copper tubing and copper pipe have been used for many years, and they are still quite popular for new installations.

Although an experienced plumber knows how to solder joints on copper pipe, sometimes the job doesn't go by the book. A little water trapped in a copper pipe can make soldering a joint a hair-pulling experience if you don't know how to handle the situation. Also at times the fittings that must be soldered are located in places where the flame from a torch poses some potentially serious problems.

Finding leaks in copper pipe and tubing is not usually difficult. By the time plumbers get called in to fix a leak, someone usually knows where the leak is. There are times, though, when the location of the leak is not known.

Big leaks are easy to find; you can either see or hear them. It is the tiny leak which does its damage over an extended period that is difficult to put your finger on. With these leaks it is sometimes necessary to start at the evidence of the leak and work your way back to the origin of the water. This can mean cutting out walls and ceilings, but sometimes there just isn't any other way to pinpoint the problem.

Copper leaks often spray water in many directions. This can also make finding the leak difficult. If a solder joint weakens to the point that water can spray out of it, the water may travel quite a distance before it splashes down. These leaks are not hard to find when they are exposed; but if they are concealed, you can find yourself several feet away from the leak when you cut into a wall or ceiling that is showing water damage.

The good thing about spraying leaks is that you can usually hear them once you have made an access hole. This is a big advantage compared to trying to find a mysterious drainage leak. The spraying water will lead you to the leak in a hurry.

Copper pipe and the joints made on it can leak for a number of reasons. Pinholes in the pipe can result from acidic water. The pipe can swell and split or blow out of a fitting if it has frozen. Stress can break joints loose, and sometimes joints that were never soldered properly will blow completely out of a fitting. Bad solder joints can also begin to drip slowly.

Fixing copper water lines is usually not a complex procedure, but what will you do if there is water in the lines? Water in the piping can be of two types: It can be standing water that is trapped, and it can be moving water that is leaking past a closed valve. In either case, the water makes it hard to get a good solder joint.

Inexperienced plumbers often don't know how to solve the problem of water in piping that they need to solder. Some inexperienced plumbers will try to make a solder joint and be fooled by the actions of the solder. This will result in a new leak, but the leak may not show up immediately; let me explain.

When water is present in the area being soldered, several things can happen. If there is enough water in the pipe, the joint will not get hot enough to melt solder. This is frustrating, but at least the plumber is aware that the solder joint can't be made without something being done about the standing water.

In some cases the pipe and fitting will get hot enough to allow solder to melt, but the fitting will not reach a temperature suitable for a solid joint. When this happens, solder will melt and roll out around the fitting, but it is not being sucked into the fitting as it should be. To inexperienced eyes, this type of joint can look okay, but it's not.

When the water is turned back on, the defective joint may leak immediately—if you're lucky. If the joint leaks right away, the plumber knows the job is not done. However, sometimes these fouled joints will not leak immediately, and this means trouble.

When the weak joint doesn't leak during the initial inspection, it may be left and possibly concealed by a wall or ceiling repair. In time—and it usually won't be long—the bad joint will begin to leak. It may drip, or it may blow out. Either way, the plumber's insurance company won't be happy.

When the bad joint fails and is inspected by the next plumber, it will be obvious that the joint was not made properly. There will not be evidence of solder deep in the fitting, because it was never there. This will usually be considered neglectful on the professional plumber's part.

A third way that water in the piping will plague an inexperienced plumber is due to steam building up in the pipe. As the area around the joint is heated, the water turns to steam. The steam vents itself, usually through some portion of the fitting being soldered. This steam may escape without being seen. When this happens, solder runs around the fitting as it should, except that a void is created where the steam is blowing out. If the plumber can't see, hear, or sense the steam, the joint will look good. Once the water is turned on, the joint will no longer look so good—it will leak, and the process will have to be repeated.

Inexperienced plumbers will think they just had a leak because of poor soldering skills. They will go back through the same process and wind up with the same results. Until they figure out what is happening, and what to do about it, they will just spin their wheels trying to solder around the steam. There are ways to solve all these problems, and I'm about to show them to you.

Trapped water

Trapped water is easier to overcome than moving water. Trapped water can sometimes be removed by bending the cut ends of horizontal pipes down. Opening fixtures above and below the work area often can be used to remove trapped water from pipes. When the problem pipe is installed vertically, a regular drinking straw can be inserted into the pipe and the water blown out. If you have the equipment and time, you can use an air compressor to blow trapped water out of pipes. When none of these options work, you have to get a bit more creative.

The easiest way to deal with trapped water in pipes that just won't drain involves the use of *bleed fittings*. These are cast fittings that are made with removable drain caps on them. Heating mechanics call them *vent fittings,* I call them bleed fittings, and many people just call them *drain fittings*.

Drain fittings are available as couplings and ells. The vent on a coupling is on the side of the fitting, at about the centerpoint. Drain ells have their weep holes on the back of the ell, where the ell makes its turn.

If you install a drain fitting in close proximity to the trapped water, you can steam the water out of the pipe as you solder the fitting. Remove the cover cap and the little black seal that covers the drain opening. Make sure the drain opening is not pointing toward you.

As you heat the pipe and fitting, the trapped water will turn to steam and vent through the weep hole in the fitting. The hot water can spit out of the hole, and the steam can come out fast and hot. This is why you don't want the hole pointed in your direction.

With the water and steam exiting the vent hole, you can solder the joint with minimal problems. Unless you have a huge amount of water in the pipe, the soldering process shouldn't take long and should go about the same as if the water weren't there in the pipe.

Once the joints have cooled, you can replace the black seal and the cap to finish your watertight joint. If you get caught without bleed fittings on your truck, a stop-and-waste valve can be substituted for the drain fitting. The little weep drain on the side of the valve will work in the same way described for the drain fittings. Make sure, however, that the valve is in its open position before you begin to solder.

Bread is an old standby for seasoned plumbers who are troubled by water in the pipes they are trying to solder. Inserting bread into the pipe will block the water long enough for a good joint to be soldered. When the water pressure is returned to the pipe, the bread will break down and come out of a faucet. This procedure works well, but there are a few warnings to be heeded.

I've used bread countless times to control water that was inhibiting my soldering. On occasion, I've regretted it. Sometimes I've created more problems for myself by using bread. I recommend drain fittings rather than bread, but I'll tell you what to look out for when bread is used, just in case you don't have a choice.

First, always remove the crust from loaf bread before stuffing the pipe. The crust is much more dense than the heart of the bread, and it doesn't dissolve as well. When bread has been used in a pipe, try to remove it through the spout of a bathtub. If you must get it out through a faucet, remove the aerator before you turn on the water. The broken-down bread will clog the screen of the aerator in the blink of an eye.

Don't flush a toilet to remove bread from the line. The

bread particles may become lodged in the fill valve and cause more trouble. Avoid packing the bread in the pipes too tightly. I once used a pencil to push and pack bread into a pipe that had moving water in it. The bread blocked the water and allowed me to make my solder joint, but it didn't come out once the water was turned on.

Moving water

Moving water makes the job of soldering joints an adventure. Old valves don't always hold back water completely. Often a trickle, or more, of water will creep past the valve and make soldering a bad experience. When there is moving water in a pipe, you will have to find some special way to do the soldering. Bleed fittings will let you do your work if there isn't too much water.

There is a special tool which I've seen advertised that is designed to made soldering wet pipes possible. As I recall, the tool has special fittings that are inserted into the pipe and expanded. The plug holds back the water while a valve is installed on the pipe. When the valve is soldered onto the end of the pipe, the tool is removed and the valve is closed, allowing the plumber to work from the valve without the threat of water.

The tool is nifty, but not needed. You can do the same thing with the old standby, bread. Stuff the pipe with bread, and pack it if you have to. Once the water is stopped, solder a gate valve onto the end of the pipe. Don't use a stop-and-waste valve, because you may need access to the bread in order to get it out.

Once the valve is soldered properly and cooled, close it and turn on the water. Open the valve to let the bread out of the system, and remember that water will be coming out right behind the bread. If the bread is packed too tightly for the water to come through, poke holes in it with a piece of wire, such as a coat hanger. The bread will break up and come out of the pipe. Once the pipe is clear, close the valve.

PB Leaks

PB leaks are not common, except at connections that were made improperly. Due to its flexibility and durability, PB pipe rarely gives plumbers problems with leaks. It doesn't even normally burst under freezing conditions. There are, however, several times when the pipe is not put together properly, and this can result in leaks.

Bad crimps account for some leaks with PB pipe. If a crimping tool gets out of adjustment—and they do after extended use—the crimp ring may not be seated properly. If this is the case, the bad connection must be removed and a new one made.

Occasionally an insert fitting is not inserted far enough. This type of problem is obvious and normally not difficult to repair. Replacement of the bad connection is all that is required.

Some plumbers are not familiar with PB pipe, and they sometimes use stainless-steel clamps to hold their connections together. This rarely works for long, if at all. If there is a leak at a PB connection where a stainless-steel clamp has been used, remove the clamp and crimp in a proper connection.

Compression ferrules may account for most of the after-installation leaks in PB piping systems. Compression fittings can be used on PB pipe, but brass ferrules should be avoided. Nylon ferrules are the proper type to use with compression fittings and PB pipe.

If a PB pipe or supply tube is leaking around the point of connection where a compression fitting has been used, inspect the ferrules. If brass ferrules have been used, replace them with nylon ferrules. The brass ferrules can cut into the PB pipe if the compression nut is tightened too much.

Polyethylene Leaks

Polyethylene (PE) leaks are not much of a consideration in water distribution systems. PE pipe is used frequently for water services, but its use is very limited in water distribution systems.

Cracked fittings and loose clamps are the most frequent causes of leaks in PE piping. The fittings can be replaced, and the clamps can be tightened or replaced. On the occasions when PE pipe develops pinhole leaks, there are a few easy ways to solve the problem. The best way is to cut out the bad section of piping and replace it. When this isn't convenient or possible, you can use repair clamps.

Repair clamps can be used to patch small holes in all types of piping. You can buy the clamps, or in some cases you make your own. For PE pipe, rubber tape and a stainless-steel pipe clamp will make a good repair on small holes. Just wrap rubber tape around the hole and clamp it into place.

Chlorinated Polyvinyl Chloride Leaks

CPVC leaks are common and, in my opinion, are a real pain to deal with. I've worked with CPVC off and on for about 20 years, and I've never liked it. The one house that I plumbed with CPVC was enough to turn me off from using it in future jobs. Since that first house, early in my career, my only association with CPVC has been in repairing it.

When you troubleshoot a water distribution system made from CPVC piping, there is a lot to look for. You have to keep your eyes open for little cracks in the pipe. Due to its makeup, CPVC will crack without a lot of provocation. These hairline cracks can be hard to find.

In addition to cracks in the piping, you have to look for cracks in the threaded fittings. Fittings that have been cross-threaded are also common in CPVC systems. And to top it all off, CPVC pipe that is not supported properly can vibrate itself to the point of weakening joints which will leak.

When you find a leak in CPVC pipe or fittings, don't attempt to reglue the fitting. Cut it out and make a good connection with new materials. Attempting to patch old CPVC will often result in frustration and continued problems.

You don't have to solder CPVC, but water in the line will still mess up new joints. The water will seep into the cement and create a void that will leak. I know that some plumbers just cut the pipe, slap a little glue on it, and stick it together with its fitting; but I don't think you should operate this way. CPVC is so finicky that I believe you should make your connections by the book.

Cut the ends of the pipe squarely, and rough them up with some sandpaper. Using a cleaner and a primer prior to applying the cement is also a good idea. When you glue and connect the pipe with the fitting, turn it, if you can, to make sure the cement gets good coverage.

Don't turn the pipe loose right away. If you release your grip too soon, the pipe is likely to push out of the fitting to some extent, weakening the joint. Hold the connection in place as long as your patience will allow. CPVC takes a long time to set up, so don't move it or turn on the water too soon. New joints should set for at least 1 h before being subject to water pressure.

Galvanized Steel and Brass

Galvanized steel pipe and brass pipe aren't used much for water distribution these days, but they still exist in some buildings. Leaking threads are a common problem with old piping. The threads are the weakest point in the piping, and they tend to be the first part of the system to go bad. Acidic and corrosive water can work on these threads to make them leak prematurely.

When there is a leak at the threads going into a fitting, a repair clamp is not going to help. Under these conditions, the leaking section of pipe must be removed and replaced. This can become quite a job. What starts out as a single leak at one fitting can quickly become a plumber's nightmare. As you cut and turn the bad section of pipe, you are likely to weaken or break the threads at some other connection. This chain reaction can go on and on until you practically have to replace an entire section of the water distribution system.

Many young plumbers see leaks at threads and believe they can correct the problem by tightening the pipe. In old piping, this is only a pipe dream (no pun intended). Twisting the old pipe tighter is not likely to solve the problem, but it may worsen it.

If there are leaks at the threads, you might as well accept that the section is going to have to be replaced. Unlike with drain pipe, you can't use rubber couplings to put water piping back together. You have to remove sections of the pipe until you can get to good threads. Then you can convert to some type of modern piping to replace the bad section. Be prepared to replace more than you plan to.

Pinhole leaks in steel and brass pipes can be repaired with repair clamps. Rubber tape and a pipe clamp will work, but a real repair clamp is best. Brass pipe imitates copper so well that some plumbers think they are working with copper. In fact, I hate to admit it, but I've even mistaken brass for copper. It is easy to do in poor lighting.

Brass water pipe can be cut with copper tubing cutters. The cut takes a little longer; but unless you suspect the pipe is brass, it can fool you. However, once you try to get a copper fitting on the end of the pipe, you'll know something is wrong; the fitting won't fit. If you think you're working with copper and can't figure out why your fittings won't work, look for an existing joint and see if it is a threaded connection. If it is, you're probably dealing with brass pipe.

If the fittings are soldered and your standard fittings won't fit the pipe, somebody plumbed the job with refrigeration tubing. When this is the case, you will need refrigeration fittings to get the job done. This one tip can save you a lot of frustration.

Compression Leaks

Compression leaks are frequent problems with water distribution systems. The compression fittings are usually easy to fix. All that is normally required is the tightening of the compression nut. These leaks occur when the nuts were not tight to begin with or when the pipe has vibrated enough to loosen the fitting. It is also possible that the connection was hit with something and loosened.

It may be necessary to remove the existing ferrule and replace it with a new one. This normally requires you to replace the section of pipe or tubing to which the old ferrule is attached. In any event, correcting compression leaks is one of the easier jobs that plumbers face.

Frozen Pipes

Frozen pipes can be a plumber's bread-and-butter money in the winter. They can also be troublesome to work with, hard to find, difficult to fix, and potentially dangerous if they are not worked with in the proper manner.

Steel pipe, and sometimes copper pipe, can be thawed with the use of a welding machine or a special pipe-thawing machine. If the leads are attached at opposite ends of the pipe, with the frozen section in the middle, the machines can produce enough warmth to thaw the pipes. I've seen plumbers do this many times over the years, but the process can be dangerous; fires can be started.

When I was a plumber in Virginia, I used to get numerous calls to thaw and repair frozen pipes. Normally, the job entailed only a single pipe, often that of an outside hose bib. A heat gun, a hair dryer, or a torch made quick work of thawing these individual sections of pipe.

In Maine, the freeze-ups are often considerably more extensive than the ones I dealt with in Virginia. Here it is not uncommon for whole systems to freeze. To thaw these pipes, I usually use a portable heater. A large space heater, like those used on construction sites, will bring a building up to a thawing temperature quickly and safely. Once the building is warm and the pipes are thawed, necessary repairs can be made.

If you haven't worked with many frozen pipes, you may not be aware of how the freezing action can swell copper pipe. Even though the split resulting from the freeze-up is in one place, the pipe might be swollen for several feet on either side of the split. This makes it impossible to get a fitting on the end of the swollen pipe. So you have to keep moving back on the pipe until you find a piece that is not swollen from the cold.

Common sense and field experience are a plumber's two best tools when troubleshooting. The more calls you handle, the more you learn. As your experience grows, the time it takes you to identify and correct a problem decreases. This chapter should shave a lot of time off your learning curve.

3

Installing
Drain-Waste-and-Vent
Systems

Installing drain-waste-and-vent (DWV) systems is a bit more complex than installing water pipes. Code regulations for drains and vents are much more numerous than those for water piping. A plumber has to know what fittings can be used to turn from a horizontal run to a vertical run. It's necessary to know the minimum pipe size for piping installed underground. How far above a roof does a vent pipe have to extend before it terminates? The answer depends on where you live, and you must be in tune with your local plumbing code to avoid getting rejection slips when your work is inspected.

Code regulations are not the only thing that complicates the installation of a DWV system. Since drains and vents are much larger in diameter than most water pipes, it is harder to find open routes where the plumbing can be installed. There is also a matter of pipe grade that must be maintained with a DWV system. This doesn't apply to water pipes.

Some plumbers prefer installing DWV systems to water systems. Others find the job of installing drains and vents to be a bother. I don't dislike either type of piping. In my opinion, water pipe is much easier to work with and install, but I enjoy running DWV pipes.

There is no question that DWV systems require planning. It's pretty easy to change directions with a water pipe, but this is not always the case with a drain or vent. Thinking ahead is crucial to the success of a job. This is true of any type of plumbing, and the rule certainly applies to DWV systems. Since many drainage systems start in the dirt, that is where we will begin—no, not in the dirt, but with underground plumbing.

Underground Plumbing

Underground plumbing is frequently overlooked in books on plumbing. Nonetheless, it is instrumental to the successful operation of some plumbing systems. Underground plumbing is installed before a house is built. When the plumbing is not installed properly, a new concrete floor will have to be broken up to relocate your pipes. This is a fast way to lose your job or make an enemy out of your customer.

Since the placement of underground plumbing is critical, you must be acutely aware of how it is installed. Miscalculating a measurement in the above-grade plumbing may mean cutting out the plumbing; but the same mistake with underground plumbing will be much more troublesome to correct.

An underground plumbing system is often called *groundwork* by professional plumbers. The groundwork is routinely installed after the footings for a home are poured and before the concrete slab is installed. When installing underground plumbing, you have to note code differences for groundworks versus above-grade plumbing. For example, while it is perfectly correct to run a 1½-in drain for an above-grade bathtub, you must run a 2-in drain if it is under concrete. Let's take a moment to look at some of the code variations encountered with interior underground plumbing.

Code considerations

Underground plumbing is quite different from above-grade plumbing. There are several rules pertaining to ground-works that do not apply to above-grade plumbing. This is true of both the DWV system and the water distribution system. The following paragraphs will highlight common differences. This information may not apply in your juris-diction, and the information does not include all code requirements. The following code considerations are the ones I encounter most often with underground plumbing.

The differences for DWV installations are minimal, but they must be observed. Galvanized steel pipe is an ap-proved material for a DWV system when the pipe is located above grade. But galvanized drain pipes may not be installed below ground. This rule has little effect on you, since most DWV systems today are plumbed with schedule 40 plastic pipe.

The minimum pipe size allowed under concrete is 2 in. This rule could cause trouble for the unsuspecting plumber. Suppose you check the sizing charts and see that $1\frac{1}{2}$-in pipe is an approved size. You might choose to use it under concrete. If you do, you will be in violation of most codes. While $1\frac{1}{2}$-in pipe is fine above grade, it is illegal below grade.

The types of fittings allowed also vary. For work above grade, a short-turn quarter bend is allowed in changing direction from horizontal to vertical. This is not true below concrete. Long-sweep quarter bends must be used for all 90° turns under the concrete.

When drains and vents are run above grade, the pipes must be supported every 4 ft. This is done with some type of pipe hanger. When installing the pipes underground, you will not use pipe hangers. Instead, you will support the pipe on firm earth or with sand or gravel. The pipe must be supported firmly and evenly.

When the groundworks leave the foundation of a house, the pipe will have to be protected by a sleeve. The sleeve can be a piece of schedule 40 plastic pipe, but it must be two pipe sizes larger than the pipe passing through the sleeve. This rule applies to any pipe passing through the foundation or under the footing.

Rarely will any residential drains under concrete run for 50 ft, but if they do, you must install a cleanout in the pipe. Pipes with a diameter of 4 in or less must be equipped with a cleanout every 50 ft in a horizontal run. The cleanouts must extend to the finished floor grade.

As the building drain becomes a sewer, there are yet more rules to adhere to. The home's sewer must have a cleanout for every 100 ft it runs. There should also be a cleanout within 5 ft of the house where the pipe makes its exit. It can usually be either inside or outside the foundation wall. These cleanouts must extend to the finished grade. If the sewer takes a turn of more than 45°, a cleanout must be installed. Cleanouts should be the same size as the pipes they serve, and they must be installed so that they will open in the direction of flow for the drainage system.

With pipes of a 3-in or larger diameter, there must be a clear space of 18 in in front of the cleanout. On smaller pipes, the clear distance must be at least 12 in. The sewer must be supported on a firm base of earth, sand, or gravel as the sewer travels the length of the trench.

Placement

The placement of groundworks is critical in the plumbing of a building. Before you install the underground plumbing, you must do some careful planning. The first step is the laying out of the plumbing.

When the groundworks are installed, the sewer and water service may or may not already be installed. If they are already installed, your job is a little easier. The pipes will be stubbed into the foundation, ready to be connected. If these pipes are installed, you can start with them and run the underground plumbing. If the water service and sewer are not installed, you will have to locate the proper spot for them.

Refer to your blueprints for the location of the water and sewer entrance. If the pipe locations are not noted on the plans, use common sense. Determine from where the water service and sewer will be coming to the house. If you are working with municipal water and sewer, the public works department of your town or city can help you. The public works department will tell you where your connections will be made to the mains and how deep they will be. If you will be connecting to a septic system and well, find the locations for these systems.

You have two considerations in picking the location of the water service and sewer. First, it should be the most convenient location inside the home for plumbing purposes. Second, there is an exit point that will allow successful connection to the main sewer and water service outside the home. Make your decision with a priority on connecting to the mains outside the home. You can adjust the interior plumbing to work for the incoming pipes, but you may not be able to adjust the existing outside conditions.

Once you have picked a location for the sewer and water service, you are ready to lay out the remainder of the groundworks. The locations for underground pipes will be partially determined by the blueprints. The blueprints show fixture locations. These fixture locations determine where you must have pipes in place. In addition to any grade-level plumbing, you look to the plumbing on upper floors. Where there are plumbing fixtures above the level of the concrete floor, you have to rough in pipes for the drains.

When you turn pipes up out of the concrete for drains and vents, location of the pipes can be crucial. Many of these pipes are intended to be inside of walls to be built. If pipe placement is off by even 1 in, the pipe may miss the wall location. When this happens, a pipe will stick up through the floor in the wrong place. It may be in a hall or a bedroom, but it will have to be moved. Moving the pipe will require the breaking and repairing of the new concrete floor. To avoid these problems, make careful measurements (Fig. 3.1) and check all measurements twice.

Figure 3.1 Measuring to the center of a drain for a groundwork installation.

Figure 3.2 A typical trenched-in groundwork installation.

When you have decided on all the pipes you will need in the underground plumbing, you can lay out your ditches. You will normally have to dig ditches to lay in the pipes. The easiest way to mark your ditches is with lime or flour. By placing these white substances on the ground, in the path of the ditch, you can dig the ditches accurately. The ditches have to be graded to enable the proper pitch of your pipes (Fig. 3.2). The standard pitch for household plumbing is $\frac{1}{4}$ in of fall for every 1 ft the pipe travels.

Installation

With all the planning and layout done, you are ready to start the installation. The best place to start is with the sewer. The height of the exit point is dictated by the depth of the connection at the main sewer. If the sewer is not already stubbed into the foundation, you have to establish the proper depth for the sewer leaving the foundation. At times, if you take the sewer out under the footing, the drain will be too low to make the final connection at the main. Before tunneling under the footing or cutting a hole in the foundation, you must determine the proper depth of the hole. This can be done by measuring the distance from the main sewer to the foundation. If the main connection is 60 ft away, the drain will drop 15 in from the time it leaves the house until it reaches the final destination. This determination is made by dividing the distance by 4, to allow for $\frac{1}{4}$ in of fall per 1 ft.

The sewer pipe should be covered by at least 12 in of dirt where it leaves the foundation. With this fact taken into consideration, the main connection must be at least 27 in deep. This is arrived at by taking 12 in of depth needed for cover and adding it to the 15 in of drop in the pipe's grade. It is best to allow a few extra inches to ensure a good connection point.

I will start the instructions, assuming the water service and sewer have not been run to the house when you start plumbing. With the depth of the hole known for the sewer, tunnel under the footing or cut a hole in the foundation wall. Install a sleeve for the sewer that is at least two pipe sizes larger than the sewer pipe. Install a cap on one end of the pipe to be used for the sewer. Extend the capped end of the sewer pipe through the sleeve and about 5 ft beyond the foundation. You are now ready to install the groundworks for all the interior plumbing.

In almost every circumstance, the pipe used for underground DWV plumbing will be schedule 40 plastic pipe. Working with this pipe is easy. The pipe can be cut with saws or roller-type cutters. Most plumbers cut the pipe with a saw. The type of saw varies. I use a hacksaw on pipe up through a 4-in diameter. A hand saw, like the one carpenters use, is effective in cutting the pipe and is easier to use on pipe over 3 in in diameter. There are saws made especially for cutting schedule 40 plastic pipe.

It is important to cut the pipe evenly. If the pipe cut is crooked, it will not seat completely in the fitting. This can cause the joint to leak. After you make a square cut on the pipe, clear the pipe of any burrs and rough pieces of plastic. You can usually do this with your hand, but the burrs and plastic can be sharp. It is safer to use a pocketknife or similar tool to remove the rough spots. The pipe and the interior of the fitting's hub must be clean and dry. Wipe these areas with a cloth if necessary, to clean and dry the surface.

Apply an approved cleaning solution to the end of the pipe and the interior of the fitting hub. Next, apply an approved primer, if required, to the pipe and fitting hub. Now apply the solvent or glue to the pipe and hub. Insert the pipe into the hub until it is seated completely in the hub. Turn the pipe one-quarter turn to ensure a good seal. This type of pipe and glue makes a quick joint. If you have made a mistake, you will have trouble removing the pipe from the fitting. If the glue sets up for long, you will have to cut the fitting off the pipe to correct any mistakes.

Many plumbers dry-fit the joints to confirm their measurements. Dry-fitting is putting the pipe and fittings together without glue to check alignment and measurements. I have done plumbing for a long time and don't dry-fit, but many plumbers prefer to dry-fit each joint. This procedure makes the job go much more slowly and can create its own problems. If you are dry-fitting the pipe to get an accurate feel for your measurements, the pipe must seat tightly in the fitting. When this is accomplished, it can be difficult to get the pipe back out of the fitting.

While I never dry-fit, I do sometimes verify my measurements with an easier method. I place the fitting beside the pipe, with the hub positioned as it will be when glued. I then take a measurement to the center of my fitting to confirm the length of the pipe. Holding the fitting beside the pipe, instead of inserting the pipe, I don't have to fight to get the pipe out of the fitting. Once a joint is made with schedule 40 plastic pipe, it should not be moved for a minute or so. It doesn't take long for the glue to set up, and you can continue working without fear of breaking the joint.

The first fitting to install on the building drain is a cleanout. The cleanout will usually be made with a wye and an eighth bend. The pipe extending from the eighth bend must extend high enough for the cleanout to be accessible above the concrete floor. Many plumbers put this cleanout in a vertical stack that serves plumbing above the first floor. The cleanout can be stopped at the finished floor level, but it must be accessible above the concrete.

After the cleanout is installed, you can go on about your business. As you run your pipes to the appropriate locations, they must be supported on solid ground or an approved fill, such as sand or gravel. Maintain an even grade on the pipes as they are installed. The minimum grade should be $\frac{1}{4}$ in for every 1 ft that the pipe runs. Don't apply too much grade to the pipe. If the pipe falls too quickly, the drains may stop up when used. The fast grade will drain the pipes of water, but leave solids behind. These solids accumulate to form a stoppage in the pipe and fittings.

Support all the fittings installed in the system. If you are installing a 3-by-2 wye, the 2-in portion will be higher above the ground than the 3-in portion. Place dirt, sand, or gravel under these fittings to support them. Don't allow

Figure 3.3 A turn-up pipe between strings that indicate wall locations.

dirt, mud, or water to get in the fittings or on the ends of the pipe. A small piece of dirt is all it takes to cause a leak in the joint.

To know where to position your pipes for fixtures, vents, and stacks, it helps to have string up where the walls (Fig. 3.3) will be installed. You should also know what the finished floor level will be. The blueprints will show wall locations, and the general contractor, concrete subcontractor, or supervisor will be able to give you the finished floor grade.

You must know the level of the finished floor so that the pipes will not be too high and wind up above the floor. Once you have the grade level for the finished floor, mark the level on the foundation. Drive stakes into the ground, and stretch a string across the area where the plumbing is being installed. The string should be positioned at the same level as the finished floor. You can use this string as a guide to monitor the height of the pipes.

Using the blueprints, mark wall locations with stakes in the ground. You can use a single string and keep your pipes turning up through the concrete to one side of the string, or you can use two strings to simulate the actual wall. Using two strings will allow you to position the pipes in the center of the wall.

Plumbing measurements are generally given from the center of a drain or vent. If your water closet is supposed to be roughed in 12 in off the back wall, the measurement is made from the edge of the wall to the center of the drain.

As you near completion on the underground plumbing, secure all the pipes that need to stay in their present positions. The pipes of greatest concern are pipes coming up through the concrete and pipes for bathtubs, showers, and floor drains. The best way to secure these pipes is with stakes. The stakes should not be made of wood. In many regions, termites may come to the house for the wood under the floor. Use steel or copper stakes. The stakes should be located on both sides of the pipe (Fig. 3.4) to ensure it is not moved by other trades.

Figure 3.4 Copper stakes securing a groundwork turn-up pipe.

Figure 3.5 A tub box.

Figure 3.6 A spacer cap on a turn-up pipe for a water closet.

Figure 3.7 A water pipe is secured to a copper stake in a groundwork setup. *Note:* Insulation will be added to protect copper from concrete.

When all the DWV pipe is installed, cap all pipes. You can use temporary plastic caps or rubber caps, but cap the pipes. You do not want foreign objects to get into the drains and clog them. Where you have a tub or a shower to be installed on the concrete floor, you will need a trap box (Fig. 3.5). This is just a box to keep concrete away from the pipe and to allow the installation of a trap when the bathing unit is set. Put a spacer cap over the pipe turned up for a toilet (Fig. 3.6) to be set on the concrete floor. This spacer will keep concrete away from the pipe so that you can install a closet flange when the time comes to set the toilet. Make sure all pipes are secured before you consider the job finished (Fig. 3.7).

Aboveground Rough-ins

When you are installing aboveground rough-ins, you don't have to dig any ditches. This is good. But you do have to drill holes. Once a building is framed and under roof, you

Pipes under 4 in in diameter	$\frac{1}{4}$ in/ft
Pipes 4 in or larger in diameter	$\frac{1}{8}$ in/ft

Figure 3.8   Minimum pitch of drainage pipe in zone 1.

Pipes under 3 in in diameter	$\frac{1}{4}$ in/ft
Pipes 3 in or larger in diameter	$\frac{1}{8}$ in/ft

Figure 3.9   Minimum pitch of drainage pipe in zone 2.

Pipes under 3 in in diameter	$\frac{1}{4}$ in
Pipes 3 to 6 in in diameter	$\frac{1}{8}$ in/ft
Pipes 8 in or larger in diameter	$\frac{1}{16}$ in/ft

Figure 3.10   Minimum pitch of drainage pipe in zone 3.

can start the installation of aboveground drains and vents. If the building has underground plumbing, you will be tying into the pipes you installed earlier. Not all houses have groundworks, so the framing stage might be your first visit to a job for installing pipes.

Houses without groundworks still require water services and sewers. Refer to the information given on these two components of a plumbing system in earlier paragraphs. Make sure that you don't set yourself up with a building sewer that will be too high to connect to a main sewer or septic tank (Figs. 3.8, 3.9, 3.10). When you are comfortable with your exit locations, you can begin to lay out and install your DWV rough-in.

Hole sizes

Hole sizes are important. Some codes require that the hole size be kept to a minimum to reduce the effect of fire spreading through a home. If you drill oversized holes, they can act as chimneys for fires in a building. Choose a drill bit that is just slightly larger than the pipe you will be installing. For 2-in pipe, a standard drill bit size is $2\frac{9}{16}$ in.

When a hole is cut for a shower drain, the hole size is determined by the shower drain. Keep the hole as small as possible, but large enough to allow the shower drain to fit in it. For a tub waste, the standard hole is 15 in long and 4 in wide. This allows adequate space for the installation of the tub waste. In some jurisdictions, after the tub waste is installed, the hole must be covered with sheet metal to eliminate the risk of a draft during a house fire.

Running pipe

When your design is made and your holes are open, you are ready to run pipe. This is where you must use your knowledge of the plumbing code. There are many rules pertaining to DWV systems. If you purchase a code book, you will see pages upon pages of rules and regulations. The task seems overwhelming with all the rules to follow. If you don't lose your cool, the job is not all that difficult. By following a few rules and basic plumbing principles, you can install DWV systems with ease.

One of the most common mistakes made in drainage piping is failure to remember to allow for the pipe's grade. When you draw the design on paper, you are not likely to think of the size of the floor joists and the grade you will need for the pipes. Since many houses have their plumbing concealed by a ceiling, it is important to keep the pipes above the ceiling level. Not many people want an ugly drain pipe hanging below the ceiling in their formal dining room. These small details of planning will make a noticeable difference in the outcome of the job.

Once you know the starting and ending points of your drains and vents, you can project the space needed for adequate grade. Generally the grade is set at $\frac{1}{4}$ in/ft for drains and vents. Drains fall downward, toward the final destination. Vents pitch upward, toward the roof of the house. With a 12-ft piece of pipe, the low end will be 3 in lower than the high end. When you are drilling through floor joists, most holes will be kept at least $1\frac{1}{2}$ in above the bottom and below the top of the joist. If you follow this rule of thumb, a 2-by-8 joist will have $4\frac{1}{2}$ in for you to work with.

A 2-by-8 has a planed width of $7\frac{1}{2}$ in. From $7\frac{1}{2}$ in you deduct 3 in for the top and bottom margin. This leaves $4\frac{1}{2}$ in for you to work with. What does all this mean to you? It means that with a 2-by-8 joist system, you can run a 2-in pipe for about 10 ft before you are in trouble. The pipe diameter takes up 2 in of the remaining $4\frac{1}{2}$ in. This leaves $2\frac{1}{2}$ in to manipulate for grade. At $\frac{1}{4}$ in/ft, you can run 10 ft with the 2-in pipe. With $1\frac{1}{2}$-in pipe, you could go 12 ft. And 3-in pipe will be restricted to a distance of about 6 ft.

Under extreme conditions, you can run farther by drilling closer to the top or the bottom of the floor joist. Before you drill any closer than 1 in to either edge, consult the carpenters. They will probably have to install headers or a small piece of steel to strengthen the joists. Many experienced plumbers have trouble with running out of space for pipes. They don't look ahead and do the math before drilling the holes. After drilling several joists, they realize they cannot get where they want to go. This results in a change in layout and a bunch of joists with holes drilled in them that cannot be used. Plan your path methodically, and you will not have these embarrassing problems.

When your holes have been drilled, running the pipe will be easy. Depending on the code you are following, every fixture must have a vent. Except for jurisdictions using a combination-waste-and-vent code, you must provide a vent for every fixture. This does not mean every fixture must have an individual back-vent. Most codes allow the use of wet vents. Using wet vents will save you time and money.

Wet vents

Wet vents (Figs. 3.11 and 3.12) are pipes that serve two purposes. They are a drain for one fixture and a vent for another. Toilets are often wet-vented with a lavatory. This

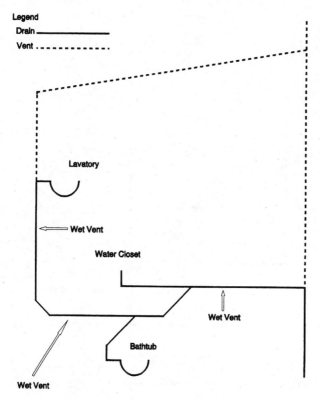

Legend
Drain _____
Vent .- - - - - - - -.

Lavatory

← Wet Vent

Water Closet

↑ Wet Vent

Bathbub

Wet Vent

Figure 3.11 Wet-venting a bathroom group.

involves placing a fitting within a prescribed distance from the toilet that serves as a drain for the lavatory. As the drain proceeds to the lavatory, it will turn into a dry vent after it extends above the trap arm. Exact distances and specifications are set forth in local plumbing codes.

Dry vents

Many of your fixtures will be vented with dry vents (Fig. 3.13). These are vents that do not receive the drainage discharge of a fixture. Since the pipes carry only air, they are called *dry vents*. There are many types of dry vents: com-

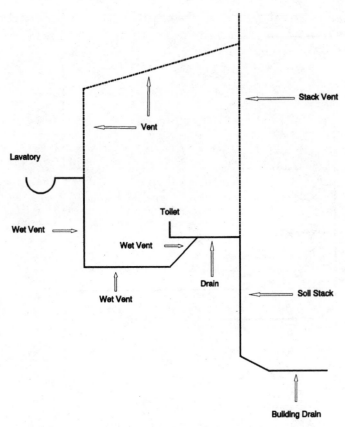

Figure 3.12 Wet-venting a toilet with a lavatory.

mon vents (Fig. 3.14), individual vents (Fig. 3.15), circuit vents (Fig. 3.16), vent stacks, and relief vents (Fig. 3.17), to name a few. Don't let all these vents scare you. In plumbing an average house, venting the drains does not have to be complicated.

If you don't understand wet vents, you can run all dry vents. This may cost a little more in material, but it can make the job easier for you to understand. Remember, every fixture needs a vent. How you vent the fixture is up

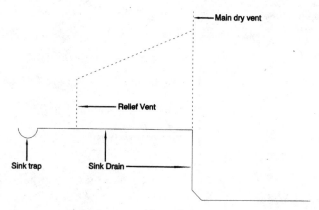

Figure 3.13 Main dry vent with a relief vent.

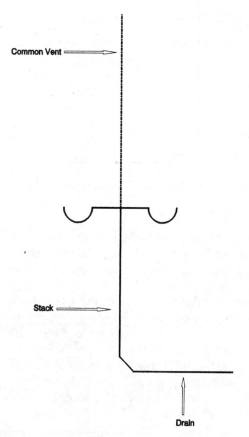

Figure 3.14 Common vent.

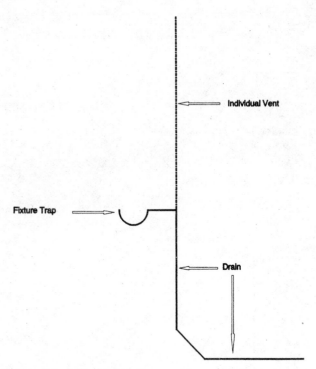

Figure 3.15 Individual vent.

to you, but as long as a legal vent is installed, you will be okay. You must install at least one 3-in vent that will penetrate the roof of the home. You can have more than one 3-in vent, but you must have *at least* one.

After you have your mandatory 3-in vent, most bathroom groups can be vented with a 2-in vent. The majority of individual fixtures can be vented with $1\frac{1}{2}$-in pipe. Most secondary vents can be tied into the main 3-in vent before the main vent leaves the attic. In a standard application, the average house will have one 3-in vent going through the roof and one $1\frac{1}{2}$-in vent, which serves the kitchen sink, going through the roof on its own.

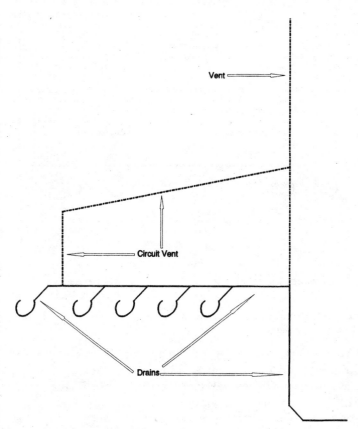

Figure 3.16 Circuit vent.

If you wanted to, you could vent each fixture with an individual vent and take them all through the roof. This would be a waste of time and money, but it would be in compliance with the plumbing code. It is more logical to tie most of the smaller vents back into the main vent, before it exits the roof. Before you can change the direction of a vent, it must be at least 6 in above the flood-level rim of the fixture it is serving.

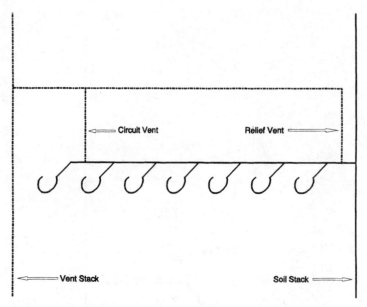

Figure 3.17 Circuit vent with a relief vent.

Yoke vents (Fig. 3.18) are not common in residential plumbing, but branch vents (Fig. 3.19), stack vents (Fig. 3.20), and vent stacks (Fig. 3.21) are. Even island vents (Fig. 3.22) are used in some residential jobs. You should become familiar with the various types of vents so that you can install efficient systems.

When you start to install the drains, you must pay attention to pipe size, pitch, pipe support, and the fittings used. The sizing charts in your local plumbing code will help you to identify the proper pipe sizes. If you grade all your pipes with the standard $\frac{1}{4}$ in/ft (Fig. 3.23), you won't have a problem there. Support all the horizontal drains at 4-ft intervals or in compliance with your local code requirements (Figs. 3.24 through 3.29). The last thing to learn about is the use of various fittings.

When each fixture in the system is vented, use P-traps for tubs, showers, sinks, lavatories, and washing machine

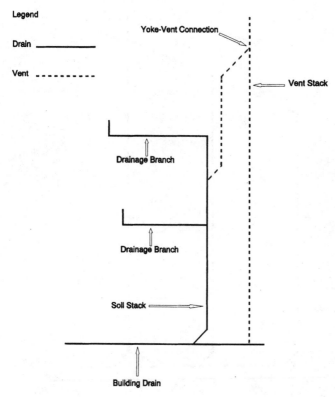

Figure 3.18 Yoke vent.

drains. When the pipes rise vertically, use a sanitary tee to make the connection between the drain, vent, and trap arm. Use long-sweep quarter bends for horizontal changes in direction. You can use short-turn quarter bends above grade when the pipe is changing from horizontal to vertical; but if you want to keep it simple, use long-sweep quarter bends for all 90° turns.

Use wyes with eighth bends to create a stack or branch that changes from horizontal to vertical. Remember to install cleanouts near the base of each stack and at the end of horizontal runs, when feasible. Keep turns in the piping

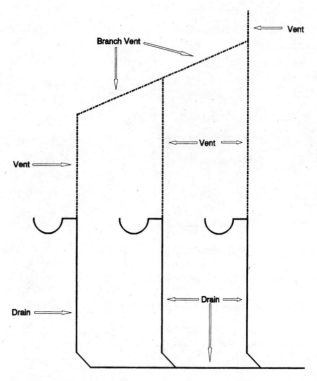

Figure 3.19 Branch vent.

to a minimum. The fewer turns the pipe makes now, the fewer problems you will have with drain stoppages later. Never tie a vent into a stack below a fixture if the stack receives that fixture's drainage. If the pipes could be hit by a nail or drywall screw, install a metal plate to protect the pipe. These nail plates are nailed or driven onto studs and floor joists to prevent damage to pipes.

Keep your vents within the maximum distance allowed from the fixture. For a 1½-in drain, the distance between the fixture's trap and the vent cannot exceed 5 ft. For 2-in pipe the distance is 8 ft for a 1½-in trap and 6 ft for a 2-in trap. A 3-in pipe has a maximum distance of 10 ft. With 4-

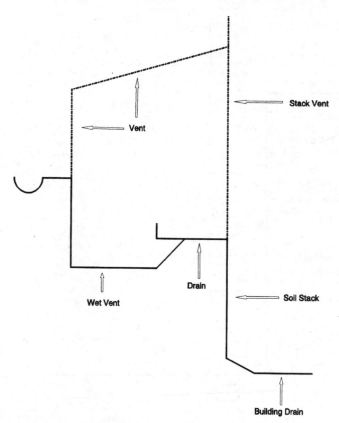

Figure 3.20 Stack vent.

in pipe, you can go 12 ft. If the fixture must be beyond these limits from the stack vent, install a relief vent.

A *relief vent* is a dry vent that comes off the horizontal drain as it goes to the fixture. The relief vent rises up at least 6 in above the flood-level rim and ties back into the stack vent.

If you are faced with a sink in an island counter, you have some creative plumbing to do for the vent. Since the sink is in an island, there will be no walls to conceal a nor-

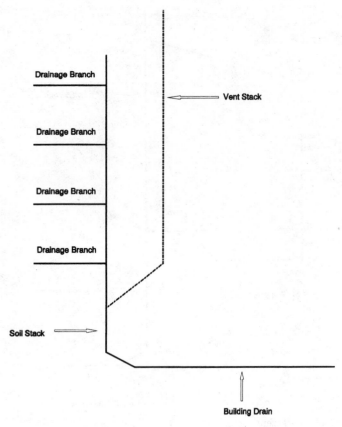

Drainage Branch

Drainage Branch

Drainage Branch

Drainage Branch

Vent Stack

Soil Stack

Building Drain

Figure 3.21   Vent stack.

mal vent. Under these circumstances you must use an *island vent*.

When your vents penetrate a roof, they must extend at least 12 in above it. In some areas, the extension requirement is 2 ft. If the roof is used for any purpose other than weather protection, the vent must rise 7 ft above the roof. Be careful taking vents through the roof when windows,

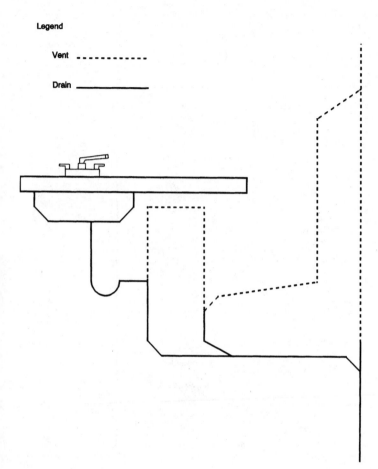

Legend

Vent ‑ ‑ ‑ ‑ ‑ ‑ ‑

Drain _____

Figure 3.22 Island vent.

doors, or ventilating openings are present. Vents must be 10 ft from these openings or 3 ft above them (Fig. 3.30).

When you are plumbing the drain for a washing machine, keep this in mind: The standpipe from the trap must be at least 18 but not more than 30 in high. The piping must be accessible for clearing stoppages.

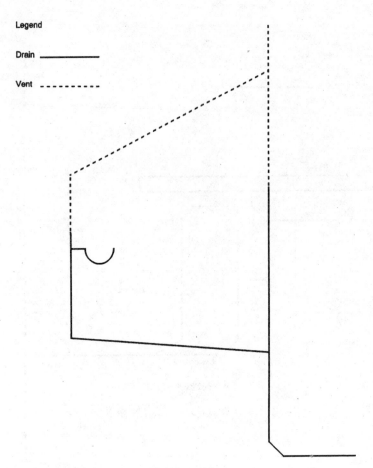

Figure 3.23   Graded-vent connection.

Fixture placement

Fixture placement is normally shown on all blueprints. Code requirements dictate elements of locating fixtures. Rough-in measurements vary from fixture to fixture and from manufacturer to manufacturer. To be safe, you should obtain rough-in books from your supplier for each type and brand of fixture you will be roughing in.

Figure 3.24   Horizontal pipe support intervals in zone 1.

Type of vent pipe	Maximum distance between supports, ft
ABS	4
Cast iron	At each pipe joint*
Galvanized	12
Copper ($1\frac{1}{2}$-in and smaller)	6
PVC	4
Copper (2-in and larger)	10

Note: Cast-iron pipe must be supported at each joint, but supports may not be more than 10 ft apart.

Figure 3.25   Horizontal pipe support intervals in zone 2.

Type of vent pipe	Maximum distance between supports, ft
ABS	4
Cast iron	At each pipe joint
Galvanized	12
PVC	4
Copper (2-in and larger)	10
Copper ($1\frac{1}{2}$-in and smaller)	6

Figure 3.26   Horizontal pipe support intervals in zone 3.

Type of vent pipe	Maximum distance between supports, ft
Lead	Continuous
Cast iron	5
Galvanized	12
Copper tube ($1\frac{1}{4}$ in)	6
Copper tube ($1\frac{1}{2}$ in and larger)	10
ABS	4
PVC	4
Brass	10
Aluminum	10

Figure 3.27   Vertical pipe support intervals in zone 1.

Type of drainage pipe	Maximum distance between supports
Lead	4 ft
Cast iron	At each story
Galvanized	At least every other story
Copper	At each story*
PVC	Not mentioned
ABS	Not mentioned

*Support intervals may not exceed 10 ft.
Note: All stacks must be supported at their bases.

Figure 3.28   Vertical pipe support intervals in zone 2.

Type of vent pipe	Maximum distance between supports
Lead	4 ft
Cast iron	At each story*
Galvanized	At each story†
Copper ($1\frac{1}{4}$ in)	4 ft
Copper ($1\frac{1}{2}$ in and larger)	At each story
PVC ($1\frac{1}{2}$ in and smaller)	4 ft
PVC (2 in and larger)	At each story
ABS ($1\frac{1}{2}$ in and smaller)	4 ft
ABS (2 in and larger)	At each story

*Support intervals may not exceed 15 ft.
†Support intervals may not exceed 30 ft.
Note: All stacks must be supported at their bases.

Figure 3.29   Vertical pipe support intervals in zone 3.

Type of vent pipe	Maximum distance between supports, ft
Lead	4
Cast iron	15
Galvanized	15
Copper tubing	10
ABS	4
PVC	4
Brass	10
Aluminum	15

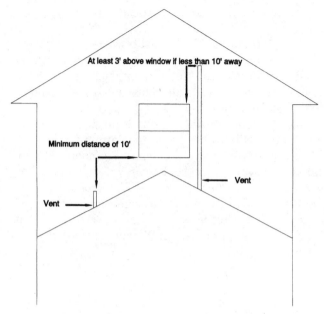

Figure 3.30 Layout of vent position to operable windows.

Standard toilets will rough in with the center of the drain 12 in from the back wall. This measurement of 12 in is figured from the finished wall. If you are measuring from a stud wall, allow for the thickness of drywall or whatever the finished wall will be. Most plumbers rough in the toilet $12\frac{1}{2}$ in from the finished wall. The extra $\frac{1}{2}$ in gives you a little breathing room if conditions are not exactly as planned. From the front rim of the toilet, there must be a clear space of 18 in between the toilet's rim and another fixture.

From the center of the toilet's drain, there must be 15 in of clearance on both sides. So you need a minimum width of 30 in to install a toilet. If you are plumbing a half-bath, the room must have a minimum width of 30 in and a minimum depth of 5 ft. These same clearance measurements apply to bidets.

If you plan to install a tub waste using slip nuts, you will need an access panel to gain access to the tub waste. If you do not want an access panel, use a tub waste with solvent-welded joints. Many people object to access panels in their hall or bedroom; avoid slip-nut connections, and you can avoid access panels. We will talk more about code requirements and placement in Chap. 22.

If you take the time to learn your local plumbing code and are willing to think ahead in your planning, DWV systems will not give you many problems. Plumbers who are intimidated by cryptic code books can refer to one of my other books, *National Plumbing Codes Handbook,* also published by McGraw-Hill, for help in sizing and other code-related issues. The book is as easy to understand as this one is, and it offers guidance for interpreting all three of the major plumbing codes.

Troubleshooting and Repairing DWV Systems

Troubleshooting and repairing drain-waste-and-vent (DWV) systems is work that can age plumbers before their time. Finding the causes of problems with DWV systems can be extremely frustrating. Even after a problem is found, figuring out how to get at it or fix it can be enough to make you want to pull your hair out. Service work is a job that many plumbers enjoy, but dealing with DWV systems can be all it takes to make a service plumber consider going into new construction.

Is working with existing DWV systems really so bad? It can be. Some of the potential problems that plumbers have to identify and solve are very aggravating. Finding a leak in a water distribution system is usually pretty easy. Since the leak is under constant pressure, it leaks all the time. This is not the case with a drain. There are a number of situations that can make troubleshooting and repairing DWV systems somewhat less than fun.

One reason so many plumbers get frustrated with DWV systems is that they have limited experience. As with most things in life, troubleshooting gets easier as experience is gained. I can't accelerate your career and give you 10 years of field experience. But I can give you the benefit of my experience, which spans more than 20 years as a plumber. When you read this chapter, keep in mind that it is not based on research or theory. The information contained here is a direct result of my hands-on experience. Let's see how much you can learn before your next service call.

Slow Drains

Slow drains, drains that empty slowly, are one of the most frequent complaints with the DWV system. Slow-moving drains can be the result of a number of possible problems. The type of drain is related to how to troubleshoot it. Knowing the type of pipe that you are working with can influence the troubleshooting steps you take. Let me give a brief rundown of the various types of pipes and the problems that they are known for.

Copper

Copper drainage systems are usually not much trouble. The copper provides many years of good service, and it is not normally a contributor to stoppages. In fact, in 20 years I can't recall a time when a copper DWV system was the cause of any problems. Oh sure, they get stopped up, but all drains can. In general, copper is a good aboveground drainage material that is very dependable. It is possible for copper drains to build up scale and slow down, but this occurs pretty rarely. Most problems with copper systems are the result of a user putting something in the drain that shouldn't be there.

Plastic pipe

Plastic pipe, as you probably know, is the most common pipe used for DWV systems today. It may be acrylonitrile butadiene styrene (ABS) or polyvinyl chloride (PVC), but either is good, and both are dependable. PVC is more brittle than

ABS and is more likely to be cracked or broken, but under normal conditions, neither pipe gives much trouble.

I have cut out sections of plastic pipe where a layer of crud was building up on the pipe walls. This type of situation could lead to a slow drain, but I've never seen a plastic pipe with a gradual buildup sufficient to render the pipe useless or even unsatisfactory. Like copper, plastic won't normally slow down or stop up unless something is put in the drain that shouldn't be there.

Galvanized pipe

Galvanized pipe has to be one of the worst materials ever used in plumbing. This pipe is famous for its ability to rust and catch every imaginable thing that goes down it. Grease and hair are especially common in stoppages in galvanized drains.

As galvanized pipes age, they develop buildups that slowly restrict their openings. Eventually, the buildup will block the pipe completely. Snaking the drain will punch holes in the obstructions, but the stoppage will recur in a matter of weeks or months.

Drains that go down very slowly and get worse as time goes on are likely to be galvanized drains. Of all the types of drain pipes in use, galvanized drains are the most likely to produce symptoms of performing more and more slowly over time.

Old galvanized drains are also known to rust out at their threads, causing leaks at the joints. The best solution to galvanized drain problems is replacement of the old piping with more modern plumbing materials. Rubber couplings make it easy to adapt new drainage materials to the old piping.

Cast-iron pipe

Cast-iron pipe has been used for DWV systems for a very long time. This pipe material has proved itself dependable. There are, however, some drawbacks to cast-iron drains.

The interior of cast-iron pipe can be rough, especially as it starts to rust. These rough surfaces catch a lot of debris as it flows down the pipes. This leads to stoppages, but they are not as bad as the ones found in galvanized pipes and they don't recur as quickly.

One of the biggest problems with cast-iron drainage systems is the removal of cleanout plugs. The old brass plugs that have been in the cleanouts for years can be next to impossible to remove.

I've seen occasions when a 24-in pipe wrench with a 2-ft cheater bar wouldn't turn the cleanout plug. If you've done much work with pipe wrenches and cheater bars, you know the kind of leverage that is being applied under these conditions. Even so, some cleanout plugs just won't budge, and others are in locations where you just can't get enough leverage on them to make the turn.

When this happens, your options are limited. Some plumbers drill out the brass plugs. Others use cold chisels and hammers to knock out the plugs. And still others cut the pipe and put it back together with a rubber coupling to make cleaning the drain the next time a bit easier. I usually opt for the last.

While cast-iron pipe will last for decades, it is susceptible to slowdowns. An electric snake equipped with a cutting head can be used to scour the pipe walls and thus solve the problem. New stoppages should not occur for years, unless there is something more wrong with the pipe, such as inadequate grading.

Bathtub Drains

Bathtub drains can run slowly for several reasons. The waste and overflow may need to be adjusted. Hair may have built up in the tub waste. The trap could be obstructed by foreign objects. The drain's vent may be obstructed. It's possible that the drain is not vented at all. These reasons, and a few others, can make a bathtub drain slowly.

If you are having trouble with a newly installed bathtub,

the problem is most likely in the tub waste and overflow. There are several types of waste and overflows. The type where you flip a lever to hold and release water in the tub may need to be adjusted. The length of the rod on these units is adjustable for tubs of various heights. If the rod has been set at a length too long for the tub, the fixture may drain slowly.

To determine whether the tub waste needs to be adjusted, remove the cover from the overflow pipe. Remove the assembly with the rod and plunger on it. Next, remove the pop-up plug from the tub's drain, if it has one. Block the tub's drain with a rag and fill the tub with water. Remove the rag or stopper and see whether the tub drains properly. If it does, adjust the tub waste. If it does not, the problem is farther along in the drainage system.

Adjusting the tub waste is simple. The rod that you removed has threads and nuts on it. The housing it screws into is usually marked with various numbers. These numbers represent the height of the tub. If you are adjusting the waste to allow water to flow more freely, shorten the rod. To do this, loosen the set nuts and screw the rod farther into the housing. You may have to find the proper setting through trial and error. Just keep adjusting the waste rod until the tub drains satisfactorily. You know that the problem is in the tub waste; all you have to do is reach the proper setting, and the tub drain will work fine.

If the tub waste does not have a trip lever, it is most likely controlled by a stopper plug. These plugs are usually depressed to open and close the tub drain. At times, the rubber gasket on the stopper may impede the flow of the draining water. When this is the case, turn the stopper counterclockwise for a few turns. With continued adjustments, you should reach a point where the drain functions properly.

Bathtub traps are usually P-traps. If you have determined that your problem is not associated with the tub waste and overflow, the trap is the next logical place to look. With new installations, there is rarely a problem with the trap, but there could be. You can run a small spring snake down the overflow pipe of the tub and into the trap. If the snake meets no resistance, the trap is not causing the problem.

If you are dealing with new plumbing, there is no reason why you should be having trouble with the bathtub's drain pipe. However, if you connected to an old piece of galvanized pipe to install your new tub, or are working on existing plumbing, the drain pipe could very well be the problem. Continue to put your snake through the trap and into the drain pipe. For some blockages, a hand snake clears the pipe. But if you have a strong blockage in a galvanized pipe, you need an electric drain cleaner to eliminate the clog.

When galvanized pipes become blocked, the stoppage often consumes the entire pipe. Small drain snakes only punch holes in the clog. The drain works better for awhile, but then it stops up again. To remove the typical stoppage from a galvanized pipe, you need an electric drain cleaner with a cutting head. As the head rotates its way through the pipe, it cuts away the debris from the sides of the pipe. Unfortunately, even a thorough cutting of the blockage is often only a temporary repair. The rusted interior of galvanized pipe will not waste time before it catches grease, hair, and other particles to create a new clog.

Shower Drains

Showers do not have a waste and overflow arrangement to adjust. When shower drains move slowly, the problem is in the trap or the drain line. The same methods described for clearing blockages in tub traps and drains apply to showers.

Sink Drains

If a sink is draining slowly, assume that the problem is in the trap or the drainage piping. Unlike tub and shower traps, the traps for most sinks are readily accessible. These traps are frequently joined with nuts that allow the disassembling of the trap. When a sink drains slowly, disassemble the trap and inspect for any obstruction that may cause the slow draining.

If the trap is clear, run a snake down the drain pipe. It is unusual for a sink drain to operate poorly following a new installation, but sink drains in older homes often fill with grease and slow down. In a lavatory there might be a problem with a pop-up assembly. The pop-up assembly may need to be adjusted to allow water to drain faster. Also, possibly, there is a hair clog on the pop-up rod. To check these possibilities, you must remove the pop-up plug.

If there is a hair clog, you will see it when you look down the drain assembly. To determine if the pop-up assembly needs adjustment, place a rag or removable stopper over the lavatory drain. Fill the lavatory bowl to capacity, and remove the rag or stopper. If the water drains well under these conditions, the pop-up assembly needs to be adjusted. If the water continues to leave the bowl slowly, your problem is in the trap or drainage pipe.

Toilet Drains

Toilets with a lazy flush can be affected by a number of afflictions. The wax seal may not be seated properly. The flush valve may not be operating properly, or the flush holes may be clogged. If all these inspections pass muster, the problem is in the drainage pipe.

Flush holes

When a toilet swirls and flushes slowly, check the flush holes. Even if a toilet is new, don't rule out the flush valve or the flush holes. The flush holes are located inside the toilet bowl, under the rim. When you flush the toilet, water should wash down the sides of the toilet bowl. If it doesn't, check the flush holes. The holes may be plugged with minerals and sediment. It is also possible that the holes were not formed properly during the making of the toilet. To check the flush holes, you need a stiff piece of wire. A wire coat hanger will work fine for checking the holes.

You can locate the holes with a mirror or by running your hand around the underside of the toilet's rim. Push the wire into the holes. If the wire goes in easily, you can rule out the flush holes as the cause of the problem. If you hit resistance, work the wire back and forth. If white, crystallized particles fall out, there is a mineral buildup. If the wire will not go in at all, it may be a defective toilet bowl.

Tank parts

If flush holes are not at fault, check out the tank parts. Inspect the water level in the tank. If the water is at the fill line, etched into the tank, then there is adequate water for a normal flush. While you have the cover off the tank, flush the toilet. Watch the flush valve when the tank empties. The flapper or tank ball should stay up long enough to allow most of the water to leave the tank. If the flapper or tank ball closes the flush valve early, there will not be enough water for a satisfactory flush. When this is the case, you must adjust the interior parts of the toilet tank.

Wax rings

If all aspects of the toilet pass inspection, you must look at the wax ring or the drain pipe. Wax rings that are not installed properly can cause a toilet to flush slowly. In new installations, wax rings may be the primary cause of lazy flushes. The problem occurs when the toilet is set on the flange. If the toilet bowl is not positioned directly over the wax ring, the wax may be compressed and may spread out over the drain pipe. Since you cannot see the wax seal when the toilet is set, the bad seal goes unnoticed.

As the toilet is used, toilet paper catches on the wax extending over the drain. After several flushes, the buildup can hinder the flushing action of the toilet. To confirm the condition of the wax ring, you must remove the toilet from its flange. As an alternative to pulling the toilet, you can use a closet auger to check the toilet's trap and drain. By running the auger through the toilet, you may retrieve wax on the end of the auger. If you do, you can bet the wax ring did not seat properly during the installation.

While the closet auger may reveal evidence of a faulty wax seal, the only way to be certain of the seal is to remove the toilet. If the hole at the bottom of the toilet has wax spread over it, you have found your problem. Scrape the old wax off and reset the toilet, using a new wax ring. Be diligent to achieve a good seal when you reset the toilet.

Drain pipes

If you have completed all these troubleshooting procedures and still have a slow-flushing toilet, then the problem is in the drain pipes. You will have to use a snake to locate and clear the blockage. During construction and remodeling, some strange objects find their way into open drain pipes. This is especially true of the drains for water closets. Since these pipes are at floor level, it is easy for foreign objects to fall down the drain.

In addition to these objects accidentally falling into the drains, sometimes someone will purposefully stick something in the pipe. When a construction job is in progress, all sorts of people have access to the job. Anyone can stuff undesirable objects in the pipes. People also often put foreign objects in toilet bowls. During my career, I have found beer cans, nails, rubber balls, blocks of wood, and countless other objects in pipes. When these objects get in the pipes, they don't go down the drain well. Sometimes the pipe may have to be cut to remove the blockage. A snake will not cut through a block of wood.

To determine if a drain pipe is stopped up, you can use a snake. Feed the snake into the drain until you come into contact with the blockage. For small drains, a spring snake will negotiate the turns in the pipe better than a flat-type snake. When you hit the blockage, try to push it through the pipe with the snake. If the blockage will not clear, try an electric drain cleaner or cut the pipe.

For drains with a diameter of 2 in or less, use a small electric drain cleaner. These units are handheld and often resemble an electric drill with a drum housing attached to it. For drains with diameters of 3 in or more, you will need a larger machine. These large machines are potentially dangerous. When a clog is encountered, these powerful machines can cause severe damage to body parts. The cable can kink and do serious damage to human appendages. Before you use an electric drain cleaner, observe proper safety precautions.

Electric drain cleaners can be rented from most rental centers. When you rent the machine, request full instructions on how to safely operate it. In general, wear gloves, don't wear loose or dangling clothes, and never allow excess cable to build up outside the drain. Slack cable can become very dangerous when you make contact with a solid blockage. Choose a flexible spring head for the drain cleaner. These heads are easier to feed down the pipe and are less likely to cause uncontrollable torque than solid cutting heads.

Frozen Drains

Drain pipes should not freeze, because they are not supposed to hold water. In the drainage system, only the traps should hold water at all times. Even though drains shouldn't freeze, they sometimes do. If drains are not working and resist the passage of a snake, you could have a frozen pipe on your hands. Obviously, this will be the case only in freezing weather.

It is difficult to diagnose a frozen drain, unless you can see the pipe. When drains are run under a home, in a basement or crawl space, you have access to the pipes. By putting your hands on the drains, you should be able to tell if they are frozen. When you cannot get to the pipes to touch them, you have to make assumptions. If the pipe could be frozen, you can try a couple of different remedies.

If the pipe is draining but is doing so slowly, then pouring very hot water down the pipe may melt the ice. If you have access to the pipe, a heat gun or a handheld hair dryer may be enough to melt the ice. When the ice is far down the pipe, you can try filling the pipe with hot water. By attaching a garden hose to the hot water connection at a washing machine, you can run hot water through the hose.

Keep an eye on the water hose. Hot water may cause the hose to overheat. Feed the hose into the drain that you suspect is frozen. Push it as far into the pipe as you reasonably can. When the hose reaches a stopping point, turn on the hot water. The hot water will force the water standing in the pipe out of the open end of the drain where the hose is inserted. As the hot water melts the ice, you can push the hose farther into the pipe. Continue this cycle until the drain is no longer clogged.

Overloaded Drains

When new plumbing has been added to an old system, the old drains may not be adequate for the increased load of drainage. For example, the existing sewer may work fine when one bathtub is draining, but may run slowly with a second tub draining at the same time. The plumbing code is designed to prevent this type of problem, but things don't always work out as they should.

If you can perform a visual inspection of all the drains, you can determine whether they are large enough to handle a specified number of fixture units. The plumbing code assigns a fixture-unit rating to each fixture. Doing some simple math will tell you whether the pipe should work properly. Even when the numbers work, the drain may not. How could this be? Well, the drain may not be working to its full capacity for many reasons.

When there are roots growing into the sewer, the open diameter of the pipe is reduced. A sewer that is partially crushed will reduce the flow of water in the pipe. If the sewer is not installed at the proper pitch, the pipe will not carry the waste as quickly as it should. All these factors could cause an old sewer to resist the demands of additional plumbing. These defects are hard to spot. You may have to expose all the old piping to discover the cause of the slow-moving drains. In plumbing, you can never be sure of anything until you can see and verify all aspects of the system.

Material Mistakes

Material mistakes account for some strange problems in the DWV systems of buildings. Good plumbers tend to think along the lines of the plumbing code when they are creating a mental picture of a plumbing problem. They basically assume the job was done to code, but this is sometimes far from the truth.

Whether you are troubleshooting a DWV system or any other plumbing problem, don't assume anything. I know it can be hard to think outside the code, but at times you must. Just as police officers have to think as criminals do in order to catch them, you have to think as an irresponsible plumber or a rank amateur sometimes, to figure out plumbing problems. Let me give you a few examples.

The Sand Trap

This first example could aptly be dubbed the *sand trap.* The story has to do with a shower drain. A friend of mine who owns his own plumbing business was in my office recently. We were discussing a project that we will be working on together, and he told me this story.

The plumber was called in to work on a shower in the basement bathroom. During the course of his work, he noticed that something was not quite right with the drain in the shower. He removed the strainer plate for a closer look and was amazed at what he saw.

Looking through the drain of the shower, he saw sand. His first thought was that the trap had somehow become filled with sand. During his work with the drain, he discovered that it turned freely in the shower base. With a little more investigation, the plumber found that the trap had not been filled with sand. It turns out that the sand was the trap.

Whoever installed the shower never bothered to connect it to a trap or a drain. The fixture simply emptied its waste water into the sand beneath the concrete slab. This is clearly a situation few plumbers would ever imagine.

The Building Drain That Wouldn't Hold Water

This next case history is about the building drain that wouldn't hold water. The job began when the customer called my office and complained of strong odors in the corner bedroom of the home. It was summer, and I suspected that the customer's problem was an overworked septic field, but I couldn't be sure without an on-site inspection.

I went to the house and met with the homeowner. She started to explain her problem to me, but I could already smell the odor, and I hadn't even gone inside the house.

The house was on a pier foundation. I walked around the left side of the home and saw two young children playing in the yard. They wore shorts and no shoes and were splashing around in a puddle. This seemed odd since it hadn't rained in days.

The closer I got to the back corner of the house, and the children, the worse the smell became. I was starting to get a feeling that made me uncomfortable.

A bathroom had recently been added to the house, near the corner bedroom. Whoever installed the plumbing did so in a way that I'd never seen before. Looking under the house, I could see the problem, but I could hardly believe what I was seeing.

The drainage hanging from the floor joists was piped with schedule 40 PVC. The PVC dropped straight down to the sewer that ran from under the house to the septic tank.

It was the sewer pipe that was creating the problem. You see, someone installed slotted drainpipe for the sewer between the house and the septic tank. Much of the sewer had never been buried in the ground. The holes were on top, but the septic tank was full, and raw sewage was seeping out of the slotted sewer pipe. Sewage was puddled under the house, under the bedroom window, and you guessed it—where the children were playing.

I asked the homeowner who had installed the plumbing, and she claimed the builder who had built the home some years back had done the work. Can you envision anyone using slotted pipe for a sewer? I couldn't either.

Duct Tape

Duct tape is a plumber's friend, and it is capable of many good uses, but it is not meant to be used as a repair clamp on a drainage line.

Many years ago I was called to a house by a homeowner who said he could hear water dripping under his home after flushing the toilet. He thought perhaps the wax seal under his toilet had gone bad.

The house was built on a crawl space foundation. My first step in troubleshooting the situation was to flush the toilet. No water seeped out around the base of the toilet, but I could hear water dripping under the home, just as the homeowner had described.

I took my flashlight and went under the home to inspect the problem. As I worked my way back to the bathroom area, I got a strong whiff of sewage. Once I was close enough to the problem to see the piping and the ground, I noticed a rather large puddle on the ground. Shining my light on the pipe, I could see the remains of duct tape wrapped around the cast-iron drain.

After a closer inspection, I found that the cast-iron pipe had a large hole in the side of it. The pipe appeared to have been hit with a hammer sometime in the past. Apparently, whoever attempted to repair the hole used duct tape to seal the pipe. Maybe this was a temporary measure until they could get materials to do the job right; I don't know.

In any event, the duct tape had long since lost its ability

to retain the contents of the pipe, and raw sewage was blowing out the side of the pipe every time the toilet was flushed.

I cut out the damaged pipe and replaced it with a section of ABS and some rubber repair couplings. The repair wasn't a big job, but it was a messy one.

When jobs are not installed with the proper materials, your troubleshooting skills are put to the test. As plumbers, we are trained to look for normal problems, not problems created by someone's use of illegal materials or installation methods.

If a drain from a lavatory is piped with 1-in PE pipe, it is difficult to understand why the pipe is stopped up or why a snake is so difficult to feed into the drain. Until you see the pipe, it is natural to assume it is of a legal size and material. You cannot, however, assume anything when you're troubleshooting plumbing.

Vent Problems

Most vent problems will appear to be drainage problems. With the exception of escaping sewer gas, vent failures will affect the drainage system. When a vent is obstructed, drains will not drain as quickly as they should. This is the most common fault found in the vent system. Vents provide air to the drains, enabling the drains to flow freely. When vents become clogged, the drains don't work as well as they should. If there is a break in the vent piping, sewer gas may enter the home. The unpleasant odor associated with sewer gas is the best indicator of a broken vent of a dry trap. Normally, these two potential failures are the only ones encountered with the venting system.

Sewer gas

Sewer gas is potentially dangerous. It may be harmful to human health when inhaled, and it is explosive when contained in unvented conditions. Sewer gas is essentially methane gas. The gas is formed in septic tanks and sewers. When the traps and vents in your home are working properly, you will not notice any sewer gas. If there is a defect in these systems, you may notice an unpleasant smell. Some people fail to identify the odor and develop health problems because of the escaping sewer gas. In small quantities, sewer gas offers minimal risk; but in large quantities, especially in confined conditions, the gas is potentially fatal.

Sewer gas can invade a home from many sources. When a trap on a fixture goes dry, sewer gas can infiltrate a dwelling. If there are holes or bad connections in vent piping, sewer gas can escape. If a vent is terminating near a window, door, or ventilating opening, sewer gas may be pulled into the home through the opening.

Clogged vents

If vents become obstructed, drains will not drain well. When vents become clogged, leaves are a prime suspect. If leaves fall into the vent pipe, they can become lodged in fittings and build up. With enough leaves in it, the vent can become useless. Occasionally, a bird will build a nest on, or in, a vent pipe. The nest blocks the flow of air and disables the vent. These two situations do not occur often; but if vents fail, check for these types of stoppages in the pipe.

Frozen vents

In cold climates, vent pipes sometimes freeze. The moisture accumulated in a vent pipe will begin to freeze on the walls of the pipe. If the temperature remains cold for an extended period, the ice will build up until it blocks the pipe. Cast iron and galvanized pipes are more likely to freeze than plastic pipe. Unless the temperatures are brutally cold for a long time, it is unlikely that vents will freeze to such an extent as to hamper the operation of a plumbing system.

Persistence and patience are both important qualities for troubleshooting DWV systems. Look for the unexpected. Never underestimate how far from compliance with the plumbing code a plumbing system may be. It is often helpful to have a helper along when you are looking for problems with DWV systems. One of you might need to be upstairs flushing a toilet while the other is under the home watching the piping. A planned troubleshooting approach is your best weapon against defects in DWV systems.

5

Faucets

Faucets are an essential part of a plumbing system. As technology has improved, faucets have become more complex than ever before. They can assess the water demand and turn themselves off. Some faucets sense the presence of a person and turn on automatically. Antiscald faucets save people from serious burns. Simple faucets can be purchased for less than $10, while fancy faucets sell for thousands of dollars. With all the diversity in faucets, it's easy to see why modern plumbers have trouble keeping up with the times.

This chapter introduces most of the faucets that you may have occasion to work with. Many of the faucets may be familiar to you, and some probably will not be. I'll share my troubleshooting experiences with various faucets with you. By the time you have finished reading this chapter, you should be well prepared to fix most any faucet.

Kitchen Faucets

Kitchen faucets come in many shapes and sizes. There are wall-mounted faucets, single-handle faucets, two-handle faucets, faucets with spray attachments, faucets without spray attachments, and specialty faucets. These faucets can have 6-in centers or 8-in centers. They can be one-piece bodies, or they can have individual components that must be put together to make a working faucet. Clearly, there are a lot of possibilities for all sorts of problems with kitchen faucets.

Faucet washers

At one time, faucet washers were found in all faucets. Times have changed, and so have the inner workings of faucets. Nowadays faucets don't always have washers or bib screws in them. Many of the modern faucets have springs, rubber seals, cartridges, and washerless stems. Gone are the days of putting an assortment of washers, screws, and seats on your truck to take care of any problem that might come along with a dripping faucet.

While the need for a lot more parts has evolved over time, the principles behind what makes a faucet leak are still pretty much the same as 10 or 20 years ago. We will get into these principles and practices in just a little while.

Washerless faucets

Washerless faucets are kind of like computers; they are great when they work as they are supposed to, and they are horrendous when they don't. In washerless faucets, it is necessary to replace entire stem units. These stems can cost nearly as much as a replacement faucet.

Cartridge-style faucets

Cartridge-style faucets are my favorite. I like these faucets because they are so simple to rebuild. It also happens that I've used one particular brand of cartridge-style faucet for about 15 years with almost no warranty callbacks, and this means a lot to me or to any other plumbing contractor. There is also a side benefit to my favorite cartridge-style faucet; if for any reason the hot water and cold water are piped to the wrong side of the faucet, the cartridge can be rotated to correct the problem. This is much easier than crossing supply tubes or pipes, and it doesn't appear as a mistake on the plumber's part.

Ball-type faucets

Ball-type faucets are very common and popular. This style of faucet is not difficult for plumbers to repair, but there are a lot of individual parts to be concerned with. Unlike cartridge-style faucets, where there is only the cartridge to replace, in ball-type faucets the ball, springs, and rubber parts can all give you trouble.

Disposable faucets

Disposable faucets are becoming more and more a part of today's plumbing. Plumbers used to fix faucets, but now they often just replace them. Some old-school plumbers say this is done because the new crop of plumbers don't have the knowledge or desire to repair faucets. I disagree with this generality. Being somewhat of an old-school plumber myself, I can understand what the other veterans are talking about, but I can also see the younger plumbers' points of view.

Many modern faucets are not worth repairing. In fact, a lot of faucets are available that cost less than a replacement stem for more expensive faucets. For example, I can buy a complete lavatory faucet, including pop-up assembly, for less than $10. When you consider that stems can run anywhere from $6 to $17 apiece, why would you repair an old faucet when it would be cost-effective to replace the entire unit? The customer gets a new faucet, and the plumber reduces the risk of a callback from a repair that doesn't last.

Now that we have touched on the basic types of faucets in use today, let's devote some time to troubleshooting and repairing the various types.

Much of what you are about to learn can be applied to more than just kitchen faucets. Many of the same troubleshooting techniques can be used for any type of faucet, and a lot of the repair options will be similar to those available for other types of faucets. Whether you are trying to eliminate a drip in a kitchen faucet, a lavatory faucet, a bar sink faucet, or a deck-mounted laundry faucet, most of the work will be closely related. Now with that in mind, let's jump right into the troubleshooting and repair of kitchen faucets.

Dripping from the spout

What's wrong when there is water dripping from the spout of a kitchen faucet? Usually, the problem lies with either the faucet seat or the faucet stem or washer. However, the problem is not always caused by these parts being defective. Are you wondering how the faucet could be dripping if the parts are not defective? You should be, and that's good; it means you are thinking. I will explain how the problem could be caused by something other than defective faucet parts in a few moments.

When you arrive to troubleshoot a dripping faucet, first you must determine what type of faucet you are dealing with. In other words, does the faucet have washers, a ball assembly, a cartridge, or some other type of mechanism? Experienced plumbers will know the answer to this question as soon as they see the faucet in many cases. If you can't tell by looking at the exterior of the faucet, it is easy enough to determine the type by removing the handle assembly; and you're going to have to do this anyway, so even rookies are not at a great disadvantage. A visual inspection of the faucet seats, washers, and/or stems will tell you what the problem is, most of the time.

An exception.   There is an exception, the one I mentioned just a few moments ago. Sometimes grit gets between the seat and the stem or washer. When this happens, the stem or washer cannot seat properly, and a drip results.

Plumbing systems where iron particles are present are most likely to experience a drip without having defective plumbing parts. Gravel, sand, and other impurities can also inhibit a faucet from closing fully. The debris in the faucet will usually be visible, but sometimes it will be small enough to avoid detection. If you suspect the leak is being caused by debris, flush out the stem hole with water. You can do this by turning on the water to the faucet slowly while the stem is removed. Water will bubble out of the hole and flush the grit out. Don't turn the water on too fast, or you will create a geyser that can shoot all the way up to a ceiling.

Faucet seats

Faucet seats sometimes become worn or pitted. When this happens, faucets drip. Small grinding wheels can be used to resurface the seats, but it is usually better to replace the damaged parts.

Corrosive elements in water, such as iron and acid, are common causes of deteriorating faucet seats. Finding such problems in a plumbing system opens the door for the potential sale of water-conditioning equipment.

Washers

Washers in faucets can wear out or be damaged by rough faucet seats. If a seat has become pitted and rough, it can cut a washer to the point where the faucet will leak. If you inspect faucet washers and find them punctured or cut, replace the faucet seats at the same time as you replace the washers.

If the bib screws that hold the washers to the stems are weak, you have found evidence of corrosive water conditions. This gives you yet another opportunity to recommend water treatment equipment to the customer.

Washerless stems

Washerless stems are not considered repairable. When these components go bad, they should be replaced. It may be possible to extend their life a short time by sanding the bottom of the stem with a light-grit sandpaper, but you cannot expect long-term success with this method.

Single-handle faucets

Single-handle faucets that are dripping are going to require some replacement parts, unless the drip is being caused by

debris lodged in the faucet. The parts could be a single cartridge, O-rings, springs, ball assemblies, or rubber seals. It
is usually worthwhile to rebuild these types of faucets completely to avoid the risk of unwanted warranty callbacks.

Leaking around the base of a spout

When water is leaking around the base of a spout from a
kitchen faucet, you can bet that the problem is a bad O-
ring. You don't need much troubleshooting ability to solve
this problem. Simply remove the spout and replace the O-
ring(s).

Leaking around a faucet handle

When water is leaking around a faucet handle, either the
packing in the stem has gone bad, or there is a gasket or O-
ring around the stem that is defective. Once you remove
the handle from the faucet, you will be able to determine
which problem you are faced with.

If water is leaking out around the base of the stem, you
have a gasket or O-ring problem. When the water is coming
out around the part of the stem that turns when the faucet
is used, the problem lies with the packing. Neither of these
problems is difficult to correct.

With the water turned off, you can remove the stem and
replace the gasket or O-ring quickly. If the problem is with
the packing, remove the stem assembly and install new
packing material. In either case, the job will take only a
few minutes to complete.

Faucet won't turn off

What should you look for when the water from the faucet
won't cut off? Well, first turn off the water supply to the
faucet. With that done, proceed with troubleshooting. What
are you likely to find? You may find a large piece of grit
between the faucet seat and the washer or stem. It is also
possible that the bib screw has deteriorated and allowed

the washer to float about, resulting in a steady stream of water that won't cut off. A quick look into the faucet body should tell you what is causing the problem.

Poor water pressure

Poor water pressure is a common complaint. When the pressure is low at only one fixture, the troubleshooting skills needed to solve the problem are not extensive. Before you begin a massive investigation into the inner workings of the troublesome faucet, there are a few simple things to check.

Start your troubleshooting by removing the aerator from the faucet. Look at the screen and diverter disk to see whether they are blocked by debris or mineral deposits. Turn on the faucet and see what the water pressure is like with the aerator removed. Many times this will prove to be the simple solution plumbers hope for. If the aerator is the culprit, all you have to do is clean it or replace it.

When the aerator is not at fault, check the cutoff valves under the fixture. Make sure the valves are in the full-open position. You might be surprised at how often the valves are partially closed and causing low water pressure at faucets.

While you are checking the cutoff valves, look at the supply tubes running up to the faucet. If a supply tube has a bad crimp in it, you may have found the cause of the low water pressure. Also see what type of material has been used for the water distribution piping.

If the water distribution piping is made of galvanized steel, you probably have your work cut out for you. Old galvanized steel water piping tends to clog with rust and mineral deposits. As the pipe clogs slowly, the water pressure to the plumbing fixtures is gradually decreased. The reduction is not noticeable in the early stages, but in advanced stages the flow can be reduced to a trickle. This type of problem requires extensive work. The old pipe must be removed and new piping installed.

Once you have eliminated the external causes of low

water pressure, you must look inside the faucet. You may find debris blocking the faucet inlets or delivery tubes. One way to be sure the problem is in the faucet is to disconnect a supply tube and turn on the water. If you have good pressure at the tip of the supply tube, the problem has to be in the body of the faucet.

It may be necessary to replace some faucets that have become clogged with mineral deposits or debris. In other cases you can simply clear the blockage and put the faucet back into service.

Nonuniform water flow

When a faucet is delivering water that does not have a uniform flow, the problem is with the aerator. A spraying or rough stream of water is a sure indication of a partially plugged aerator. Remove the aerator and clean or replace it.

No water at the spray head

Many plumbing customers complain that they have no water at the spray head of the kitchen faucet. There are only a few possible causes for this type of problem.

When you have a spray attachment that will not deliver water, check the spray hose. These hoses frequently become kinked under sinks. A bad kink can render the spray head useless.

If the spray hose is in good condition, take the spray head apart and inspect for mineral deposits or debris. You will often find iron deposits blocking up the works. If necessary, replace the spray head.

I can't recall its ever happening, but the outlet port on the bottom of the faucet could become blocked, cutting off the water supply to the spray hose. If you have replaced the old spray assembly with a new one and are still not getting water, consider the possibility of a blockage at the connection port.

Limited water at the spray head

Limited water at the spray head is more common than a complete lack of water. The same troubleshooting steps offered for solving no-water problems can be followed to solve problems with limited water at the spray head.

Water dripping from beneath a faucet

Water dripping from beneath a faucet can be frustrating. This type of problem is often misdiagnosed as a leak at the supply tube. If water leaks from the bottom of a faucet and runs through the faucet holes in the sink, it can appear to be coming from the supply-tube connection.

Acidic water is a primary cause of leaks in the bodies of faucets. The thin copper tubing used to mix and deliver water in the bodies of faucets can be eaten up by water with a high acid content. Small pinholes can develop and allow water to escape within the faucet housing. You won't be able to spot these leaks without first removing the entire faucet from the sink.

When a faucet body is pitted or corroded, replace it. While it is sometimes possible to repair the leak with a spot-soldering job, it is better to replace the faucet.

Lavatory Faucets

Lavatory faucets are very similar to kitchen faucets when it comes to troubleshooting and repair work. There are, however, more multipiece lavatory faucets used than there are kitchen faucets. Multipiece faucets are faucets that consist of individual handles and spouts that are connected, by plumbers, with tubing beneath the fixture. While single-body faucets are much more common, multipiece faucets are frequently found in more expensive plumbing systems. These faucets offer a few features that single-body faucets don't.

Most of the troubleshooting techniques used for multi-piece faucets are the same as those used for single-body faucets. However, the connection points under fixtures for multipiece faucets can produce leaks that don't exist in single-body faucets.

Many multipiece faucets are connected with small copper tubing and compression fittings. The tubing is soldered into the individual faucet components and is connected to the mixing spout with compression fittings. The tubing is so small and thin that it crimps easily. This can result in low water pressure. It is also easy to break the soldered connections between the tubing and the individual components.

When you are working with multipiece faucets, be careful. Too much stress on the components can twist, crimp, or break the tubing connections. Since multipiece faucets are typically expensive, you don't want to have to replace one at your own expense.

In the interest of space and time, I will not give a blow-by-blow account of how to work with lavatory faucets. You can use the same principles and techniques described for kitchen faucets to work with lavatory faucets.

Bar Sinks and Laundry Tubs

Deck-mounted faucets for bar sinks and laundry tubs fall into the same troubleshooting and repair category as kitchen faucets. As for lavatory faucets, refer to the section on kitchen faucets for information on troubleshooting the faucets for bar sinks and laundry tubs.

Bathtubs and Showers

The faucet valves for bathtubs and showers are a little different from those used for kitchen sinks. The tools and techniques required for troubleshooting these valves are similar, but not always the same. Let's see how the differences affect your work.

Dripping from the spout

When water is dripping from a tub spout or a showerhead, you can use the same basic principles described for kitchen faucets in your troubleshooting and repair work. You may, however, need a set of tub wrenches to remove the valve stems. Tub wrenches, for those of you who aren't familiar with them, resemble deep-set sockets.

Two- and three-handle tub and shower valves are often equipped with valves whose wrench flats are concealed in the finished wall. Since you can't get a standard wrench on the flats to remove the stems, you must have tub wrenches to reach into the wall and remove the parts.

Once you have removed the stems for tub and shower valves, the rest of the troubleshooting and repair procedure is comparable to that used for kitchen faucets.

Leaking around a faucet handle

When water is leaking around a faucet handle on a tub or shower valve, the cause of the leak is usually defective packing around the stem. The stems for tub and shower valves are larger than those found in sink faucets, but they share many of the same characteristics. By removing the stem you can replace the bad packing. It is possible that the leak is coming from a bad O-ring, so check this possibility before reinstalling the stem.

Valve won't cut off

If you have a tub or shower valve that won't cut off, you can apply the same basic principles and procedures used for kitchen faucets in your troubleshooting and repair work.

Poor water pressure

Poor water pressure is rare with bathtub valves, but showerheads sometimes suffer from a lack of desirable water pressure. The cause is usually a mineral buildup in the water-saver disk that is housed in the showerhead. If you remove the head and check the disk for obstructions, you will probably find some. A quick rinse under a faucet will often clear the debris. In stubborn cases you may have to poke the pieces out with a thin piece of wire or replace the disk. Also the holes in the showerhead itself may be plugged.

Lack of uniform flow

Lack of a uniform flow is another problem you could encounter with a showerhead that is partially plugged with foreign objects. While tub spouts won't give you this type of trouble, showerheads can. When the spray from a showerhead is erratic, check the head and water-saver disk for blockages.

Diverters that don't divert well

Diverters that don't divert well are a fairly common problem with three-handle tub-and-shower valves. Sometimes water isn't diverted from the tub spout to the showerhead. At other times water runs out of both the tub spout and the showerhead simultaneously. The cause of these types of problems is the stem washer or the valve seat. Treat the condition just as you would a dripping faucet, and replace the seat or washer as needed.

Wall-Mounted Faucets

Wall-mounted faucets are not very different from their deck-mounted cousins when it comes to troubleshooting and repairing them. You can apply the same principles and procedures discussed for kitchen faucets to your work with wall-mounted faucets.

Frost-free Hose Bibs

Frost-free hose bibs are similar to a sink faucet in terms of troubleshooting. The stem for a frost-free hose bib is, however, much longer than a stem for a standard faucet. The fixture still contains a seat and a washer, but the stem is extra long. The length of the stem depends on the length of the hose bib, but it can easily be up to 1 ft long.

Hose bibs also have packing and packing nuts that can spring leaks. You can apply the basic troubleshooting principles and practices we discussed in the section on kitchen faucets to frost-free hose bibs.

Learn the Basics

Once you learn the basics of troubleshooting valves and faucets, you will be able to tackle just about any type of unit with a good deal of success. The general principles are the same for all types of faucets and valves. The job can get tricky only when you fail to follow a systematic troubleshooting approach.

As routine as troubleshooting faucets is, it can be frustrating. There will no doubt be times when the problem doesn't know it is supposed to go by the book. When this happens, you must depend on the experience and knowledge you have gained through field work and study to solve the problem. To drive this point home, I take a little time now to look at some scenarios that you might run into out in the field.

Field Conditions

Field conditions can play a major role in the level of difficulty a plumber faces in solving on-the-job problems. Repair work that should be simple can become quite complicated when there are space limitations. For example, a multipiece faucet that should be simple to repair can become a nightmare if the faucet is offset to one corner of a sink and drawers in the cabinet block access. The same could be said for trying to replace a washer when the bib screw has been eaten up by acidic water.

The act of replacing a faucet washer is certainly not complicated. Under normal conditions, most people are capable of handling the work. If, however, the head of the bib screw falls apart when you touch it with a screwdriver, that simple job turns into a challenging assignment. If this were to happen to you, what would you do?

When the head of a bib screw disintegrates, you have a few options. The least complicated one is replacement of the faucet stem. Some customers, however, will not want to pay for a whole new stem. If customers knew how much they were going to spend for the time it takes a plumber to work with a broken screw, they would be more inclined to replace the stem—and this is something you may want to mention to cost-conscious customers.

If the customer insists on having you spend time working with the broken screw, accept the challenge and overcome it. There are two fairly simple ways of doing this. You can take a knife and cut out the old washer. This will expose the threaded portion of the old screw. Many times the shaft of the screw will not have rotted so far that it will not turn. If you put pliers on the threaded shaft, once the washer is removed, there is a good chance you can turn the shaft out of the stem.

Should the shaft of the screw break off, don't panic. Use a small drill bit to drill out the old screw. Then either you can try to rethread the stem, or you can install snap-in washers. Snap-in washers are fitted with two ears that can be pushed into the hole you drilled. Once the ears are in the hole, their natural tension will hold the washer in place.

Every experienced plumber will have a personal way to deal with a broken bib screw. What is the right way to get the job done? Any approach that works is acceptable. You should strive for a procedure that is fast, cost-effective, dependable, and durable and that satisfies the customer; when you meet these criteria, you have done a fine job.

Field conditions can send some plumbers into tailspins. If the plumbers have not spent a considerable time doing real plumbing, they can be mystified by the actions of other plumbers, homeowners, and even inanimate plumbing.

This type of situation can arise around any type of plumbing, not just faucets and valves.

I've been on countless jobs where the field conditions made doing my job all but impossible. Even now, with 20 years of experience, I still encounter circumstances in the field that test my abilities to the limits. For example, what kind of a person would install a water heater with the element access doors facing a solid wall? Obviously, someone who had no intention of ever working on the water heater again. I saw an installation like this just last week.

Why would someone thread a coupling onto the supply-tube inlets of a faucet to connect the supply tube ? One of my recent jobs had such an arrangement, and there was a leak between the faucet inlet and the supply tube. If the coupling had not been installed, it would not have been possible for such a leak to exist. The coupling was not needed, and it was not standard plumbing procedure to install it. To make matters worse, the coupling had been installed so tightly that it took a basin wrench with a 14-in pipe wrench attached to its handle to remove the couplings.

What would inexperienced plumbers think when seeing this strange coupling in a place where the books never show such a fitting? They would probably be confused, and they might try to salvage the fitting or replace it. With my experience I knew the coupling was useless, so I removed it and installed a slightly longer supply tube. The leak was fixed.

The two most important traits to develop when working in the field are *experience* and *product knowledge.* Experience comes only with time, but product knowledge can be learned as soon as you are willing to make the effort. If you know how plumbing fixtures, faucets, and devices are put together, you are well on your way to being an excellent troubleshooter. On the other hand, if you are not aware that a particular faucet should have a gasket in a particular place, how will you know what the problem is if the gasket is missing?

All major plumbing manufacturers are very willing to

provide professionals with detailed information on products. If you take the time to obtain cut sheets from manufacturers, you can begin your study of what makes faucets and valves work. This same approach applies to all plumbing parts, devices, and fixtures. Just ask your supplier for the information you want. If the supplier doesn't have detailed drawings, you can get the manufacturer's address from the supplier and request the information directly from the company that makes the plumbing parts. Fixing faucets is not difficult when you have the right tools and information to work with.

6

Fixtures

When we talk of plumbing fixtures, there are numerous types of fixtures which might be a part of the conversation. They range from a simple hose bib to a bidet. Plumbing codes dictate what fixtures must be installed in various types of buildings. The code sets minimum standards, but of course, other fixtures can be added to a building. So choosing the types of fixtures to install in a building is one consideration. But there is a lot more to fixtures than just which ones are required. I warn you, this will be a long chapter, but it is needed, because the subject of fixtures is broad.

What Fixtures Are Required?

The number and type of fixtures required will depend on local regulations and the use of the building where they are being installed. Your code book will tell you what is required and how many of each type of fixture are needed. These requirements are based on the use of the building housing the fixtures and the number of people who may be using the building. Let me give you some examples.

Single-family residence

When you are planning fixtures for a single-family residence, you must include certain ones. If you choose to
install more than the minimum, that's fine, but you must
install the minimum number of required fixtures. The minimum number and type of fixtures for a single-family
dwelling are as follows:

- One toilet
- One bathing unit
- One lavatory
- One kitchen sink
- One hookup for a clothes washer

Multifamily buildings

The minimum requirements for a multifamily building are
the same as those for a single-family dwelling, except that
each dwelling in the multifamily building must be
equipped with the minimum fixtures. There is one exception—the laundry hookup. With a multifamily building,
laundry hookups are not required in each dwelling unit. In
zone 3, it is required that a laundry hookup be installed for
common use when the number of dwelling units is 20. For
each interval of 20 units, you must install a laundry
hookup.

For example, in a building with 40 apartments, you have
to provide two laundry hookups. If the building had 60
units, you would need three hookups. In zone 1, the
dwelling-unit interval is 10 rental units. In zone 2 one hookup is required for every 12 rental units, but no less than two
hookups are needed for buildings with at least 15 units.

Nightclubs

For businesses and places of public assembly, such as
nightclubs, the ratings are based on the number of people

likely to use the facilities. In a nightclub, the minimum requirements for zone 3 are as follows:

- *Toilets*—one toilet for every 40 people
- *Lavatories*—one lavatory for every 75 people
- *Service sinks*—one service sink
- *Drinking fountains*—one drinking fountain for every 500 people
- *Bathing units*—none

Day care facilities

The minimum number of fixtures for a day care facility in zone 3 is as follows:

- *Toilets*—one toilet for every 15 people
- *Lavatories*—one lavatory for every 15 people
- *Bathing units*—one bathing unit for every 15 people
- *Service sinks*—one service sink
- *Drinking fountains*—one drinking fountain for every 100 people

In contrast, in zone 2 only toilets and lavatories must be installed in day care facilities. The ratings for these two fixtures are the same as in zone 3, but the other fixtures required in zone 3 are not required in zone 2. This type of rating system is found in your local code and covers all the normal types of building uses.

In many cases, facilities will have to be provided in separate bathrooms, to accommodate both sexes. When separate bathroom facilities are installed, the number of required fixtures is divided equally between the two sexes, unless there are cause and approval for a different appropriation.

Some types of buildings do not require separate facilities. For example, in zone 3 the following buildings are not required to have separate facilities: residential properties, small businesses, where less than 15 employees work, or where less than 15 people are allowed in the building at the same time.

In zone 2 separate facilities are not required in the following buildings: offices with less than 1200 square feet (ft^2), retail stores with less than 1500 ft^2, restaurants with less than 500 ft^2, self-serve laundries with less than 1400 ft^2, and hair salons with less than 900 ft^2.

Employee and customer facilities

There are some special regulations pertaining to employee and customer facilities. For employees, toilet facilities must be available within a reasonable distance and with relative ease of access. For example, in zone 3 these facilities must be in the immediate work area; the distance that an employee is required to walk to the facilities may not exceed 500 ft. The facilities must be located such that employees do not have to negotiate more than one set of stairs for access to the facilities. There are some exceptions to these regulations, but it general these are the rules.

It is expected that customers of restaurants, stores, and places of public assembly shall have toilet facilities. In zone 3, this is based on buildings capable of holding 150 or more people. Buildings in zone 3 with an occupancy rating of less than 150 people are not required to provide toilet facilities, unless the building serves food or beverages. When facilities are required, they may be placed in individual buildings, or in a shopping mall situation, in a common area, not more than 500 ft from any store or tenant space. These central toilets must be placed so that customers will not have to use more than one set of stairs to reach them.

In zone 2 a square-footage method is used to determine minimum requirements in public places. For example, retail stores are rated as having an occupancy load of 1 person for every 200 ft^2 of floor space. This type of facility is required to have separate facilities when the store's area exceeds 1500 ft^2. A minimum of one toilet is required for each facility when the occupancy load is up to 35 people. One lavatory is required in each facility, for up to 15 people. A drinking fountain is required for occupancy loads up to 100 people.

Handicap Fixtures

Handicap fixtures are not cheap; you cannot afford to overlook them when bidding on a job. The plumbing code will normally list specific minimum requirements for handicap-accessible fixtures in certain circumstances. It is your responsibility to know when handicap facilities are required. There are also special regulations pertaining to how handicap fixtures are to be installed. We are about to embark on a journey into handicap fixtures and their requirements.

When you are dealing with handicap plumbing, you must consult both the local plumbing code and the local building code. These two codes work together in establishing the minimum requirements for handicap plumbing facilities. When you step into the field of handicap plumbing, you must play by a different set of rules. Handicap plumbing is like a different code, all unto its own.

Where are handicap fixtures required?

Most buildings frequented by the public are required to have handicap-accessible plumbing fixtures. The following handicap examples are based on zone 3 requirements. The plumbing codes in zones 1 and 2 do not go into as much detail on handicap requirements.

Single-family homes and most residential multifamily dwellings are exempt from handicap requirements. A rule-of-thumb standard for most public buildings is the inclusion of one toilet and one lavatory for handicap use.

Hotels, motels, inns, and the like are required to provide a toilet, lavatory, bathing unit, and kitchen sink, where applicable, for handicap use. Drinking fountains may also be required. This provision will depend on both the local plumbing code and the local building code. In plumbing a gang shower arrangement, such as in a school gym, at least one of the shower units must be handicap-accessible. Door sizes and other building code requirements must be observed for handicap facilities. There are local exceptions to these rules; check with your local code officers for current, local regulations.

When it comes to installing handicap plumbing facilities, you must pay attention to the plumbing code and the building code. In most cases, approved blueprints will indicate the requirements of your job, but in rural areas, you may not enjoy the benefit of highly detailed plans and specifications. When it comes time for a final inspection, the plumbing must pass muster along with the open space around the fixtures. If the plumbing fails inspection, your pay is held up and you are likely to incur unexpected costs. This section will apprise you of what you may need to know. It is not all plumbing, but it is all needed information when you are working with handicap facilities.

When you think of installing a handicap toilet, you probably think of a toilet that sits high off the floor. But do you think of the grab bars and partition dimensions required around the toilet? Some plumbers don't, but they should. The door to a privacy stall for a handicap toilet must have a minimum of 32 in of clear space.

The distance between the front of the toilet and the closed door must be at least 48 in. It is mandatory that the door open outward, away from the toilet. Think about it. How could a person in a wheelchair close the door if the door opened in to the toilet? These facts may not seem like your problem, but if your inspection doesn't pass, you don't get paid.

The width of a water closet compartment for handicap toilets must be a minimum of 5 ft. The length of the privacy stall must be at least 56 in for wall-mounted toilets and 59 in for floor-mounted models. Unlike regular toilets, which require a rough-in of 15 in to the center of the drain from a sidewall, handicap toilets require the rough-in to be at least 18 in off the sidewall.

Then there are the required grab bars. Sure, you may know that grab bars are required, but do you know the mounting requirements for the bars? Two bars are needed for each handicap toilet. One bar should be mounted on the back wall, and the other will be installed on the sidewall. The bar mounted on the back wall must be at least 3 ft long. The first mounting bracket of the bar must be mounted no more than 6 in from the sidewall. Then the bar must extend at least 24 in past the center of the toilet's drain.

The bar mounted on the sidewall must be at least 42 in long. The bar should be mounted level and with the first mounting bracket located no more than 1 ft from the back wall. The bar must be mounted on the sidewall that is closest to the toilet. This bar must extend to a point at least 54 in from the back wall. If you do the math, you will see that a 42-in bar is pushing the limits on both ends. A longer bar will allow more assurance of meeting the minimum requirements.

When a lavatory is to be installed in the same toilet compartment, the lavatory must be installed on the back wall. The lavatory must be installed such that its closest point to the toilet is no less than 18 in from the center of the toilet's drain. When a privacy stall of this size and design is not suitable, there is another option.

Another way to size the compartment to house a handicap toilet and lavatory is available. At times space restraints will not allow a 5-ft-wide stall. In these cases, you may position the fixture differently and use a stall with a width of only 3 ft. In these situations, the width of the privacy stall may not exceed 4 ft.

The depth of the compartment must be at least 66 in when wall-mounted toilets are used. The depth extends to a minimum of 69 in with the use of a floor-mounted water closet. The toilet requires a minimum distance of 18 in from the sidewalls to the center of the toilet drain. If the compartment is more than 3 ft wide, grab bars are required, with the same installation methods as described before.

It the stall is created at the minimum width of 3 ft, then grab bars, with a minimum length of 42 in, are required on each side of the toilet. These bars must be mounted no more than 1 ft from the back wall, and they must extend a minimum of 54 in from the back wall. If a privacy stall is not used, the sidewall clearances and the grab bar requirements are the same as listed in these two examples. To determine which set of rules to use, you must assess the shape of the room when no stall is present.

If the room is laid out much like the first example, use the guidelines for grab bars listed there. If, however, the room tends to meet the description of the second example, use the specifications in that example. In both cases, the door to the room may not swing in to the toilet area.

Design of handicap fixtures

Handicap fixtures are specially designed for people with different physical abilities from the general public. The differences in handicap fixtures may appear subtle, but they are important. Let's look at the requirements of a handicap fixture.

Toilets. Toilets will have a normal appearance, but they will sit higher above the floor than a standard toilet. A handicap toilet will rise to a height of 16 to 20 in off the finished floor. For most handicap toilets 18 in is a common height. There are many choices in toilet style:

- Siphon jet
- Siphon wash
- Siphon vortex
- Reverse trap
- Blowout

Sinks and lavatories. Visually, handicap sinks, lavatories, and faucets may appear to be standard fixtures, but their method of installation is regulated and the faucets are often unlike standard faucets. Handicap sinks and lavatories must be positioned to allow a person in a wheelchair to use them easily.

The clearance requirements for a lavatory are numerous. There must be at least 30 in of clearance in front of the lavatory. This clearance must extend 30 in from the front edge of the lavatory or countertop, whichever protrudes farther, and to the sides. If you can set a square box 30 in by 30 in in front of the lavatory or countertop, you have adequate clearance for the first requirement. This applies to kitchen sinks and lavatories.

The next requirement calls for the top of the lavatory to be no more than 35 in from the finished floor. For a kitchen sink, the maximum height is 34 in. Then there is the knee clearance to consider. The minimum allowable knee clearance requires 29 in in height and 8 in in depth. This is measured from the face of the fixture, lavatory, or kitchen sink. Toe clearance is another issue. A space 9 in high and 9 in deep is required, as a minimum, for toe space. The last requirement deals with hot water pipes. Any exposed hot water pipes must be insulated, or shielded in a way, to prevent users of the fixture from being burned.

Sink and lavatory faucets. Handicap faucets frequently have blade handles. The faucets must be located no more than 25 in from the front edge of the lavatory or counter, whichever is closer to the user. The faucets could use wing handles, single handles, or pushbuttons to be operated, but the operational force required by the user shall not be more than 5 pounds (lb).

Bathing units. Handicap bathtubs and showers must meet the requirements of approved fixtures, as any other fixture, but they also require special features and installation methods. The special features are required under the code for approved handicap fixtures. The clear space in front of a bathing unit must be a minimum of 1440 in². This is achieved by leaving an open space of 30 in in front of the unit and 48 in to the sides. If the bathing unit is not accessible from the side, the minimum clearance is increased to an area 48 in by 48 in.

Handicap bathtubs are required to be installed with seats and grab bars. A grab bar, for handicap use, must have a diameter of at least $1\frac{1}{4}$ in. The diameter may not exceed $1\frac{1}{2}$ in. All handicap grab bars are meant to be installed $1\frac{1}{2}$ in from walls. The design and strength for these bars are set forth in the building codes.

The seat may be an integral part of the bathtub, or it may be a removable, after-market seat. The grab bars must be at least 2 ft long. Two of these grab bars are to be mounted on the back wall, one over the other. The bars are to run horizontally. The lowest grab bar must be mounted 9 in above the flood-level rim of the tub. The top grab bar must be mounted a minimum of 33 in but no more than 36 in above the finished floor. The grab bars should be mounted near the seat of the bathing unit.

Additional grab bars are required at each end of the tub. These bars should be mounted horizontally and at the same height as the highest grab bar on the back wall. The bar over the faucet must be at least 2 ft long. The bar on the other end of the tub may be as short as 1 ft.

The faucets in these bathing units must be located below the grab bars. The faucets used with a handicap bathtub must be able to operate with a maximum force of 5 lb. A personal, handheld shower is required in all handicap bathtubs. The hose for the handheld shower must be at least 5 ft long.

Two types of shower stalls are normally used for handicap purposes. The first type allows the user to leave a wheelchair and shower while sitting on a seat. The second type is designed for the user who rolls a wheelchair into the stall and showers while seated in the wheelchair.

If the shower is intended to be used with a shower seat, its dimensions should form a square, with 3 ft of clearance. The seat should be no more than 16 in wide and mounted along the sidewall. This seat should run the full length of the shower. The height of the seat should be between 17 and 19 in above the finished floor. Two grab bars must be installed in the shower.

These bars should be located between 33 and 36 in above the finished floor. The bars are intended to be mounted in an L shape. One bar should be 36 in long and run the length of the seat, mounted horizontally. The other bar should be installed on the sidewall of the shower. This bar should be at least 18 in long.

The faucet for this type of shower must be mounted on the wall across from the seat. The faucet must be at least 38 in but not more than 48 in above the finished floor. There must be a handheld shower. The handheld shower can be in addition to a fixed showerhead; but there must be a handheld shower on a hose at least 5 ft long. The faucet must be able to operate with a maximum force of 5 lb.

Drinking units. The distribution of water from a water cooler or drinking fountain must occur at a maximum height of 36 in above the finished floor. The outlet for drinking water must be located at the front of the unit, and the water must flow upward for a minimum distance of 4 in. Levers or buttons to control the operation of the drinking unit may be mounted on the front of the unit or on the side, near the front.

Clearance requirements call for an open space of 30 in in front of the unit and 48 in to the sides. Knee and toe clearances are the same as required for sinks and lavatories. If the unit is made so that the drinking spout extends beyond the basic body of the unit, the width clearance may be reduced from 48 to 30 in, as long as the knee and toe requirements are met.

Standard Fixture Installation Regulations

Standard fixtures must also be installed according to local code regulations. There are space limitations, clearance requirements, and predetermined, approved methods for installing standard plumbing fixtures. First, let's look at the space and clearance requirements for common fixtures.

Standard fixture placement

Toilets and bidets require a minimum distance of 15 in (Fig. 6.1) from the center of the fixture's drain to the nearest sidewall. These fixtures must have at least 15 in of clear space between the center of their drains and any obstruction, such as a wall, cabinet, or other fixture. With this rule in mind, a toilet or bidet must be centered in a space of at least 30 in (Fig. 6.2). Further, in zone 1 there

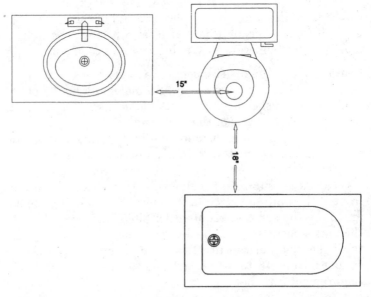

Figure 6.1 Minimum distances for legal layout.

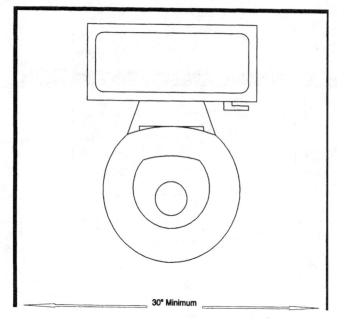

Figure 6.2 Minimum width requirements for a water closet.

must be a minimum of 18 in of clear space in front of these fixtures, and when toilets are placed in privacy stalls, the stalls must be at least 30 in wide and 60 in deep.

In zones 1 and 2 urinals must be installed with a minimum clear distance of 12 in from the center of the drains to the nearest obstacle on either side. When urinals are installed side by side in zones 1 and 2, the distance between the centers of their drains must be at least 24 in.

In zone 3 urinals must have minimum sidewall clearances of at least 15 in. In zone 3, the center-to-center distance is a minimum of 30 in. Urinals in zone 3 must also have a minimum clearance of 18 in in front of them.

These fixtures, as with all fixtures, must be installed level and with good workmanship. The fixture should normally be set with an equal distance from walls, to avoid a crooked or cocked appearance (Figs. 6.3 and 6.4). All fix-

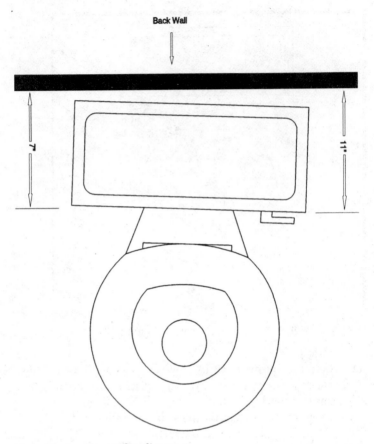

Figure 6.3 Improper toilet alignment.

tures should be designed and installed with proper cleaning in mind. Bathtubs, showers, vanities, and lavatories
should be placed in a manner to avoid violating the clearance requirements for toilets, urinals, and bidets.

Securing and sealing fixtures

Some fixtures hang on walls, and others sit on floors. When
securing fixtures to walls and floors, you must obey certain

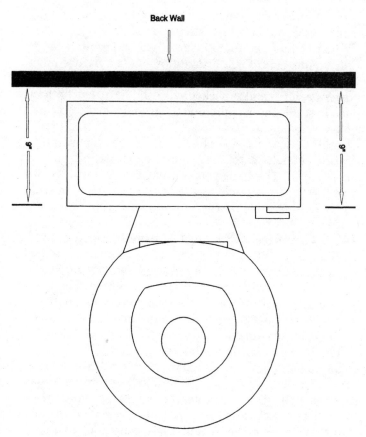

Figure 6.4 Proper toilet alignment.

rules. Floor-mounted fixtures, like most residential toilets, should be secured to the floor by a closet flange. The flange is first screwed or bolted to the floor. A wax seal is then placed on the flange, and closet bolts are placed in slots, on both sides of the flange. Then the toilet is set into place. The closet bolts should be made of brass or some other material that will resist corrosive action. The closet bolts are tightened until the toilet does not move from side to side or from front to back. In some cases, a flange is not used. When a flange is not used, the toilet should be secured with corrosion-resistant lag bolts.

When toilets or other fixtures are being mounted on a wall, the procedure is a little different. The fixture must be installed on, and supported by, an approved hanger. These hangers are normally packed with the fixture. The hanger must assume the weight placed in and on the fixture, to avoid stress on the fixture itself.

In the case of a wall-hung toilet, the hanger usually has a pattern of bolts extending from the hanger to a point outside the wall. The hanger is concealed in the wall cavity. A watertight joint is made at the point of connection, usually with a gasket ring, and the wall-hung toilet is bolted to the hanger.

With lavatories, the hanger is usually mounted on the outside surface of the finished wall. A piece of wood blocking is typically installed in the wall cavity to serve as a solid surface on which the bracket is mounted. The bracket is normally secured to the blocking with lag bolts. The hanger is put in place, and lag bolts are screwed through the bracket and finished wall, into the wood blocking. Then the lavatory is hung on the bracket.

The space where the lavatory meets the finished wall must be sealed. This is true of all fixtures coming into contact with walls, floor, or cabinets. The crevice where the fixture meets the finished surface must be sealed to protect against water damage. A caulking compound, such as silicone, is normally used for this purpose. This seal does more than prevent water damage. It eliminates hard-to-clean areas and makes the plumbing easier to keep free of dirt and germs.

When bathtubs are installed, they must be installed level, and they must be properly supported. The support for most one-piece units is the floor. These units are made to be set into place, leveled, and secured. Other types of tubs, such as cast-iron tubs, require more support than the floor will give. They need a ledger or support blocks placed under the rim, where the edge of the tub meets the back wall.

The ledger can be a piece of wood, such as a wall stud. The ledger should be about the same length as the tub. This ledger is installed horizontally and level. The ledger should be at a height that will support the tub in a level fashion or with a slight incline, so that excess water on the rim of the tub will run back into the tub. The ledger is nailed to the wall studs.

If blocks are used, they are cut to a height that will put the bathtub into the proper position. Then the blocks are placed at the two ends (and often in the middle) of where the tub will sit. The blocks should be installed vertically and nailed to the stud wall.

When the tub is set into place, the rim at the back wall rests on the blocks or ledger, for additional support. This type of tub has feet on the bottom, so that the floor supports most of the weight. The edges where the tub meets the walls must be caulked. If shower doors are installed on a bathtub or shower, they must meet safety requirements set forth in the building codes.

Showers today are usually one-piece units. These units are meant to be set in place, to be leveled, and to be secured to the wall. One-piece showers and bathtubs are secured by placing nails or screws through a nailing flange, which is molded as part of the unit, into the stud walls. If only a shower base is being installed, it must be level and secure.

Now, let's look at some of the many other regulations involved in installing plumbing fixtures.

More facts about fixture installations

When it is time to install fixtures, many rules and regulations govern. Water supply is one issue. Access is another. Air gaps and overflows are factors. There are a host of requirements governing the installation of plumbing fixtures. We will start with the fixtures most likely to be found in residential homes. Then we will look at the fixtures normally associated with commercial applications.

Typical residential fixture installation

Typical residential fixture installations could include everything from hose bibs to bidets. This section considers each fixture that could be a typical residential fixture and tells you more about how it must be installed.

With most plumbing fixtures, water is coming into the fixture and going out of the fixture. The incoming water lines must be protected against freezing and back-siphonage. Freeze protection is usually ensured by the placement of the piping. In cold climates it is advisable to avoid locating pipes in outside walls. Insulation is often applied to water lines to reduce the risk of freezing. Back-siphonage is typically avoided with the use of airgaps and backflow preventers.

Some fixtures, such as lavatories and bathtubs, are equipped with overflow routes. These overflow paths must be designed and installed to prevent water from remaining in the overflow after the fixture is drained. They must also be installed so that back-siphonage cannot occur. This usually means nothing more than having the faucet installed so that it is not submerged in water if the fixture floods. By keeping the faucet spout above the high-water mark, you create an air gap. The path of a fixture's overflow must carry the overflowing water into the trap of the fixture. This should be done by having the overflow path follow the same pipe that drains the fixture.

Bathtubs must be equipped with wastes and overflows. In zones 1 and 3 these wastes and overflows must have a minimum diameter of $1\frac{1}{2}$ in. The method for blocking the waste opening must be approved. Common methods for holding water in a tub include

- Plunger-style stoppers
- Lift-and-turn stoppers
- Rubber stoppers
- Push-and-pull stoppers

Some fixtures such as handheld showers pose special problems. Since the shower is on a long hose, it could be dropped into a bathtub full of water. If a vacuum formed in the water pipe while the showerhead was submerged, the unsanitary water from the bathtub could be pulled back into the potable water supply. This is avoided with the use of an approved backflow preventer.

When a drainage connection is made with removable connections, such as slip nuts and washers, the connection must be accessible. This normally is not a problem for sinks and lavatories, but it can create some problems with bathtubs. Many builders and home buyers despise having an ugly access panel in the wall where their tub waste is located. To eliminate the need for this type of access, the tub waste can be connected with permanent joints. This could mean soldering a brass tub waste or gluing a plastic one. But if the tub waste is connected with slip nuts, an access panel is required.

Washing machines generally get their incoming water from boiler drains or laundry faucets. There is a high risk of a cross-connection when these devices are used with an automatic clothes washer. This type of connection must be protected against back-siphonage. The drainage from a washing machine must be handled by an indirect waste. An air break is required and is usually achieved by placing the washer's discharge hose into a 2-in pipe as an indirect waste receptor. The water supply to a bidet must also be protected against back-siphonage.

Dishwashers are another likely source of back-siphonage. These appliances must be equipped with either a backflow protector or an air gap, installed on the water supply piping. The drainage from dishwashers is handled differently in each zone.

In zone 1 there must be an air gap on the drainage of a dishwasher. These air gaps are normally mounted on the countertop or in the rim of the kitchen sink. The air gap forces the waste discharge of the dishwasher through open air and down a separate discharge hose. This eliminates the possibility of back-siphonage or a backup from the drainage system into the dishwasher.

In zone 2 dishwasher drainage must be separately trapped and vented or discharged indirectly into a properly trapped and vented fixture.

In zone 3 the discharge hose from a dishwasher may enter the drainage system in several ways. It may be individually trapped. It may discharge into a trapped fixture. The discharge hose could be connected to a wye tailpiece in the kitchen sink drainage. Further, it may be connected to the waste connection provided on many garbage disposals.

While we are on the subject of garbage disposals, be advised that they require a drain of at least $1\frac{1}{2}$ in and must be trapped. It may seem to go without saying, but garbage disposals must have a water source. This doesn't mean you have to pipe a water supply to the disposal; a kitchen faucet provides adequate water supply to satisfy the code.

Floor drains must have a minimum diameter of 2 in. Remember, piping run under a floor may never be smaller than 2 in in diameter. Floor drains must be trapped, usually must be vented, and must be equipped with removable strainers. It is necessary to install floor drains so that the removable strainer is readily accessible.

Laundry trays are required to have $1\frac{1}{2}$-in drains. These drains should be equipped with cross-bars or a strainer. Laundry trays may act as indirect waste receptors for clothes washers. In the case of a multiple-bowl laundry tray, the use of continuous waste is acceptable.

Lavatories must have drains of at least $1\frac{1}{4}$ in in diameter. The drain must be equipped with some device to prevent foreign objects from entering the drain. These devices could include pop-up assemblies, cross-bars, or strainers.

When you install a shower, it is necessary to secure the pipe serving the showerhead with water. This riser is normally secured with a drop-ear ell and screws. It is, however, acceptable to secure the pipe with a pipe clamp.

When we talk of showers here, we are speaking only of

showers, not tub-shower combinations. The frequent use of tub-shower combinations confuses many people. A shower has different requirements from a tub-shower combination. A shower drain must have a diameter of at least 2 in. The reason is simple. In a tub-shower combination, a $1\frac{1}{2}$-in drain is sufficient because the walls of the bathtub will retain water until the smaller drain can remove it. A shower doesn't have high retaining walls; therefore, a larger drain is needed to clear the shower base of water more quickly. Shower drains must have removable strainers. The strainers should have a diameter of at least 3 in.

In zone 3, all shower bases must have a minimum area of 900 square in (in^2). This area must not be less than 30 in in any direction. These measurements must be taken at the top of the threshold, and they must be interior measurements. A shower advertised as a 30-in shower may not meet code requirements. If the measurements are taken from the outside dimensions, the stall will not pass. There is one exception to the above rule. Square showers with a rough-in of 32 in may be allowed. But the exterior of the base may not measure less than $31\frac{1}{2}$ in.

In zone 1, the minimum interior area of a shower base must be at least 1024 in^2. When you determine the size of the shower base, take the measurements from a height equal to the top of the threshold. The minimum size requirements must be maintained for a vertical height equal to 70 in above the drain. The only objects allowed to protrude into this space are grab bars, faucets, and showerheads.

The waterproof wall enclosure of a shower, or a tub-shower combination, must extend from the finished floor to a height of no less than 6 ft. Another specification for these enclosures is that they must extend at least 70 in above the height of the drain opening. The enclosure walls must be at the higher of the two determining factors. This might come into play, for example, in a deck-mounted bathing unit. With a tub mounted in an elevated platform, an enclosure that extends 6 ft above the finished floor might not meet the criterion of being 70 in above the drain opening.

Although not as common as they once were, built-up shower stalls are still popular in high-end housing. These stalls typically use a concrete base, covered with tile. You may never install one of these classic shower bases, but you need to know how, just in case. These bases are often referred to as *shower pans*. Cement is poured into the pan to create a base for ceramic tile.

Before these pans can be formed, attention must be paid to the surface under the pan. The subfloor, or other supporting surface, must be smooth and able to accommodate the weight of the shower. When the substructure is satisfactory, you are ready to make your shower pan.

Shower pans must be made of a waterproof material. In the old days, these pans were made of lead or copper. Today, they are generally made with coated papers or vinyl materials. These flexible materials make the job much easier. When a shower pan is formed, the edges of the pan material must extend at least 2 in above the height of the threshold. In zone 1 the material must extend at least 3 in above the threshold. The pan material must also be securely attached to the stud walls.

In zone 1, the shower threshold must be 1 in lower than the other sides of the shower base, but the threshold must never be lower than 2 in. The threshold must also never be higher than 9 in. When installed for handicap facilities, the threshold may be eliminated.

In zone 1 the shower base must slope toward the drain with a minimum pitch of $\frac{1}{4}$ in/ft but not more than $\frac{1}{2}$ in/ft. The opening into the shower must be large enough to accept a shower door with minimum dimensions of 22 in.

The drains for this type of shower base are new to many young plumbers. Plumbers who have worked under my supervision have attempted to use standard shower drains for these types of bases. You cannot do that because the pan will leak. This type of shower base requires a drain that is similar to some floor drains.

The drain must be installed in a way that will not allow water, which might collect in the pan, to seep around the drain and down the exterior of the pipe. Any water entering the pan must go down the drain. The proper drain has a flange that sits beneath the pan material. The pan material is cut to allow water into the drain. Then another part of the drain is placed over the pan material and is bolted to the bottom flange. The compression of the top piece and the bottom flange, with the pan material wedged between them, will create a watertight seal. Then the strainer portion of the drain screws into the bottom flange housing. Since the strainer is on a threaded extension, it can be screwed up or down, to accommodate the level of the finished shower pan.

Sinks must have drains with a minimum diameter of $1\frac{1}{2}$ in. Strainers or cross-bars are required in the sink drain. If you look, you will see that basket strainers have the basket part, as a strainer, and cross-bars below the basket. This blocks the entry of foreign objects, even when the basket is removed. If a sink is equipped with a garbage disposal, the drain opening in the sink must have a diameter of at least $3\frac{1}{2}$ in.

Toilets installed in zone 3 are required to be water-saver models. The older models, which use 5 gal per flush, are no longer allowed in zone 3 for new installations.

The seat on a residential water closet must be smooth and sized for the type of water closet it is serving. This usually means that the seat will have a round front.

The fill valve or ball cock for toilets must be of the antisiphon variety. There are still older ball cocks being sold that are not the antisiphon style. The fact that these units are available does not make them acceptable. Don't use them; you will be jeopardizing your license and yourself.

Toilets of the flush tank type must be equipped with overflow tubes. These overflow tubes do double duty as refill conduits. The overflow tube must be large enough to accommodate the maximum water intake of the water closet at any given time.

Whirlpool tubs must be installed as recommended by the manufacturer. All whirlpool tubs must be installed to allow access to the unit's pump. The pump's drain should be pitched to allow the pump to empty its volume of water when the whirlpool is drained. The whirlpool pump should be positioned above the fixture's trap.

All plumbing faucets and valves using both hot and cold water must be piped in a uniform manner. This calls for the hot water to be piped to the left side of the faucet or valve. Cold water should be piped to the right side of the faucet or valve. This uniformity reduces the risk of burns due to hot water.

In zone 3, valves or faucets used for showers must be designed to provide protection from scalding. This means that any valve or faucet used in a shower must be pressure-balanced or must contain a thermostatic mixing valve. The temperature control must not allow the water temperature to exceed 110°F. This ensures the safety of people, especially the elderly and the very young, from scalding injuries in the shower. In zones 1 and 2, these temperature control valves are not required in residential dwellings. In zone 1, when temperature control shower valves are required, the maximum allowable temperature is 120°F.

Commercial fixture applications

Drinking fountains are a common fixture in commercial applications. Restaurants use garbages disposals that are so big that they can take two plumbers to move them. Gang showers are not uncommon in school gyms and health clubs. Urinals are another common commercial fixture. And then there are water closets. Water closets are found in homes, but the ones installed for commercial applications often differ from residential toilets. Special fixtures and applications exist for some unusual plumbing fixtures, such as baptismal pools in churches. This section is going to take you into the commercial field and show you how plumbing needs vary from residential uses to commercial applications.

Let's start with drinking fountains and water coolers.

The main fact to remember about water coolers and fountains is this: They are not allowed in toilet facilities. You may not install a water fountain in a room that contains a water closet. If the building for which a plumbing diagram is being drawn will serve water, such as a restaurant, or if the building will provide access to bottled water, then drinking fountains and water coolers may not be required.

Commercial garbage disposals can be big. These monster grinding machines require a drain with a diameter of no less than 2 in. Commercial disposals must have their own drainage piping and trap. As with residential disposals, commercial disposals must have a cold water source. In zone 2, the water source must be an automatic type. These large disposals may not be connected to a grease interceptor.

Garbage can washers are not something you will find in the average home, but they are not uncommon in commercial applications. Due to the nature of this fixture, the water supply to the fixture must be protected against back-siphonage. This can be done with either a backflow preventer or an air gap. The waste pipes from these fixtures must have individual traps. The receptor that collects the residue from the garbage can washer must be equipped with a removable strainer, capable of preventing the entrance of large particles into the sanitary drainage system.

Special fixtures are just that—special. Fixtures that might fall into this category include church baptismal pools, swimming pools, fish ponds, and other such arrangements. The water pipes to any of these special fixtures must be protected against back-siphonage.

Showers for commercial or public use can be very different from those found in a residence. It is not unusual for showers in commercial-grade plumbing to be gang showers. This amounts to one large shower enclosure with many showerheads and shower valves. In gang showers, the shower floor must be properly graded toward the shower drain(s). The floor must be graded so as to prevent water at one shower station from passing through the floor area of another shower station.

The methods employed to divert water from each shower station to a drain are chosen by the designer, but it is imperative that water used by one occupant not pass into another bather's space. In zone 1 the gutters of gang showers must have rounded corners. These gutters must have a minimum slope toward the drains of 2 percent. The drains in the gutter must not be more than 8 ft from the sidewalls and not more than 16 ft apart.

Urinals are not a common household item, but they are typical fixtures in public toilet facilities. The amount of water used by a urinal in a single flush should be limited to a maximum of $1\frac{1}{2}$ gal. Water supplies to urinals must be protected from backflow. Only one urinal may be flushed by a single-flush valve. When urinals are used, they must not take the place of more than one-half of the water closets normally required. Public-use urinals must have a water trap seal that is visible and unobstructed by strainers.

Floor and wall conditions around urinals are another factor to be considered. These areas must be waterproof and smooth. They must be easy to clean, and they may not be made from an absorbent material. In zone 3, these materials are required around a urinal in several directions. They must extend to at least 1 ft on each side of the urinal. This measurement is taken from the outside edge of the fixture. The material is required to extend from the finished floor to a point 4 ft off the finished floor. The floor under a urinal must be made of this same type of material, and the material must extend to a point at least 1 ft in front of the farthest portion of the urinal.

The requirements of commercial-grade water closets vary from residential requirements. The toilets used in public facilities must have elongated bowls. These bowls must be equipped with elongated seats. Further, the seats must be hinged, and they must have open or split fronts.

Flush valves are used almost exclusively with commercial-grade fixtures. They are used on water closets, urinals, and some special sinks. If a fixture depends on trap siphonage to empty itself, it must be equipped with a flush valve or a properly rated flush tank. These valves or tanks are required for each fixture in use.

Flush valves must be equipped with vacuum breakers that are accessible. Flush valves in zone 3 must be rated as water-conserving valves. These valves must be able to be regulated for water pressure, and they must open and close fully. If water pressure is not sufficient to operate a flush valve, other measures, such as a flush tank, must be incorporated into the design. All manually operated flush tanks should be controlled by an automatic filler, designed to refill the flush tank after each use. The automatic filler is equipped to cut itself off when the trap seal is replenished and the flush tank is full. If a flush tank is designed to flush automatically, the filler device is controlled by a timer.

Troubleshooting Fixtures

Frost-free sill cocks

Frost-free sill cocks do not require much attention. If sediment gets between the seat and the washer, the sill cock may drip. The packing nut on the sill cock could be loose and leaking, but these are the only two problems you are likely to encounter. If the packing nut is leaking, just tighten it. Unless the nut is cracked or the packing is defective, tightening will be all that is necessary.

If water is dripping out of the sill cock, remove the stem. Cut off the water to the pipe supplying the sill cock. At the handle of the sill cock you will see a nut holding the stem into the body. Loosen the nut, and the stem will come out. Inspect the washer on the end of the stem. If the washer is good, leave it alone. If it is cut or worn, replace it. While the stem is out, turn on the water to the pipe supplying the sill cock. Water will blow out the end of the body, flushing any sediment and debris from the seat. After a few moments, cut off the water. Put the stem back into the body, and tighten the retainer nut. Turn on the water again, and inspect the results of your work. This procedure should stop the drip.

Washing machine hookups

Washing machine hookups may be nothing more than two boiler drains, or they may consist of a faucet device. In either case, there is not much call for adjustments. The boiler drains have a packing nut that may leak. They may also drip from a bad washer or debris between the washer and the seat. To correct either of these problems, follow the directions given in the section on sill cocks.

If hoses connect to a faucet, there will be little you can do if it is defective. This type of faucet has very few repairable parts. In most cases, if the faucet is bad, it must be replaced. If you have problems with water pressure, check the screens in the ends of the hoses for the washing machine. The cone-shaped screens in each hose filter sediment from the water pipes. It is not uncommon for these screens to become clogged (Fig. 6.5).

Figure 6.5 Troubleshooting laundry tubs.

Symptom	Probable cause
Faucet drips from spout	Bad washers or cartridge Bad faucet seats
Faucet leaks at base of spout	Bad O-ring
Faucet will not shut off	Bad washers or cartridge Bad faucet seats
Poor water pressure	Partially closed valve Clogged aerator Not enough water pressure Blockage in faucet Partially frozen pipe
No water	Closed valve Broken pipe Frozen pipe
Drains slowly	Partial obstruction in drain or trap
Will not drain	Blocked drain or trap
Gurgles as it drains	Partial drainage blockage Partial blockage in the vent
Won't hold water	Bad basket strainer Bad putty seal on drain

Tub wastes

If a tub is not holding water (Figs. 6.6 and 6.7) or is not draining properly, you must adjust the tub waste. The most common style of tub waste is the trip lever. To adjust these tub wastes, remove the two large screws that hold the cover plate on the overflow fitting. When the screws are removed, grasp the plate and pull the assembly up out of the overflow tube. If the tub is not holding water, lengthen this assembly. If the tub is not draining, shorten the assembly.

To adjust the assembly, loosen the nuts on the threaded portion of the assembly. When the nuts are loose, you can turn the threaded rod to adjust the length of the assembly.

Figure 6.6 Troubleshooting bathtubs.

Symptom	Probable cause
Won't drain	Clogged drain
	Clogged tub waste
	Clogged trap
Drains slowly	Hair in tub waste
	Partial drainage blockage
Won't hold water	Tub waste needs adjusting
Won't release water	Tub waste needs adjusting
Gurgles as it drains	Partial drainage blockage
	Partial blockage in vent
Water drips from spout	Bad faucet washers or cartridge
	Bad faucet seats
Water comes out spout and shower at the same time	Bad diverter washer
	Bad diverter seat
	Bad diverter
Faucet will not shut off	Bad washers or cartridge
	Bad faucet seats
Poor water pressure	Partially closed valve
	Not enough water pressure
	Blockage in faucet
	Partially frozen pipe
No water	Closed valve
	Broken pipe
	Frozen pipe

Figure 6.7   Troubleshooting showers.

Symptoms	Probable cause
Won't drain	Clogged drain
	Clogged strainer
	Clogged trap
Drains slowly	Hair in strainer
	Partial drainage blockage
Gurgles as it drains	Partial drainage blockage
	Partial blockage in vent
Water drips from showerhead	Bad faucet washers or cartridge
	Bad faucet seats
Faucet will not shut off	Bad washers or cartridge
	Bad faucet seats
Poor water pressure	Partially closed valve
	Not enough water pressure
	Blockage in faucet
	Partially frozen pipe
No water	Closed valve
	Broken pipe
	Frozen pipe

Find the proper setting by trial and error. Make the adjustment, reinstall the assembly, and test the drain. Continue this process until you have perfected the setting.

If tub waste is controlled by the stopper in the tub, there will be no assembly in the overflow tube. If the tub waste operates by pushing down on the stopper, the only adjustment you can make is to loosen or tighten the threads of the stopper. There is a rubber gasket on these stoppers. By turning the stopper, you can raise or lower the level of the rubber gasket. If you must lift and turn your stopper to operate the tub waste, there is no adjustment you can make to the tub waste.

Lavatories

The pop-up assembly in a lavatory may need to be adjusted (Fig. 6.8). This is done under the lavatory bowl. Check the

Figure 6.8   Troubleshooting lavatories.

Symptoms	Probable cause
Faucet drips from spout	Bad washers or cartridge Bad faucet seats
Faucet leaks at base of spout	Bad O-ring
Faucet will not shut off	Bad washers or cartridge Bad faucet seats
Poor water pressure	Partially closed valve Clogged aerator Not enough water pressure Blockage in faucet Partially frozen pipe
No water	Closed valve Broken pipe Frozen pipe
Drains slowly	Hair on pop-up assembly Partial obstruction in drain or trap Pop-up needs to be adjusted Pop-up is defective
Gurgles as it drains	Partial drainage blockage Partial blockage in vent
Won't hold water	Pop-up needs adjusting Bad putty seal on drain

knurled nut that holds the ball and rod in the drain assembly. If the nut is not tight, tighten it. If you need more adjustment, adjust the lift rod. You can do this where the lift rod attaches to the pop-up rod or where it attaches to the lift assembly. By making the lift rod shorter, where it protrudes above the faucet, you can gain more leverage to seal the drain. This is the procedure to use if the bowl is not holding water. If the bowl is not draining fast enough, allow more rod to stick up above the faucet. The extra length will push the rod down farther and will cause the bowl to drain more quickly.

If you fill a lavatory with water only to have it slowly drain out, you may have a putty problem. When the pop-up assembly seems to be properly adjusted, but water still leaks out, there may not be a good seal around the drain. This type of leak can be hard to locate. Since the leaking water leaks into the drain, it doesn't show up under the bowl. To remedy the problem, you must remove the drain assembly and reinstall it.

The drainage connections under a lavatory are often made with slip nuts and washers. It is not uncommon for these connections to develop small leaks in the first few weeks of operation. If you notice water dripping from the slip nut, tightening the nut should solve the problem.

Compression fittings often need to be tightened after they have seen some use. Until the joints have settled in, they may leak owing to pipe vibrations or movement. When a compression fitting is leaking, tightening the compression nut usually corrects the leak. If the leak is a bad one, you may have to replace the compression sleeve or nut. Sometimes the nuts crack and cause water to spray about. In either event, compression leaks are easy to fix.

Toilets

Often toilets require some adjustments. The adjustments needed (Fig. 6.9) will usually be in the toilet tank. Ball cocks are one of the items most often adjusted. There are two common types. The first has a horizontal float rod with a float or ball on the end of it. The second has a plastic float that moves up and down the shaft of the ball cock. The ball cock controls the amount of water that enters the toilet tank. Ball cocks can be adjusted to increase or reduce the amount of water in the tank. Ideally, the tank's water level should be level with the fill line etched onto the interior of the tank.

The float-rod ball cock is very common and is easily adjusted. If more water is needed in the tank, bend the float rod down. By bending the rod down, more water is needed to float the ball high enough to cut off the water. If less water is needed in the task, bend the float rod up. This

Figure 6.9 Troubleshooting toilets.

Symptoms	Probable cause
Will not flush	No water in tank
	Stoppage in drainage system
Flushes poorly	Clogged flush holes
	Flapper or tank ball is not staying open long enough
	Not enough water in tank
	Partial drain blockage
	Defective handle
	Bad connection between handle and flush valve
	Clogged vent
Water droplets covering tank	Condensation
Tank fills slowly	Defective ball cock
	Obstructed supply pipe
	Low water pressure
	Partially closed valve
	Partially frozen pipe
Will not drain	Blocked drain or trap
Makes unusual noises when flushed	Defective ball cock
Water runs constantly	Bad flapper or tank ball
	Bad ball cock
	Float rod needs adjusting
	Float filled with water
	Ball cock needs adjusting
	Pitted flush valve
	Undiscovered leak
	Cracked overflow tube
Water seeps from base of toilet	Bad wax ring
	Cracked toilet bowl
Water dripping from tank	Condensation
	Bad tank-to-bowl gasket
	Bad tank-to-bowl bolts
	Cracked tank
	Flush valve nut loose
No water comes into the tank	Closed valve
	Defective ball cock
	Frozen pipe
	Broken pipe

causes the ball to reach its cutoff point earlier, resulting in less water. The float rods bend easily in your hands.

To bend the rod, place the palm of your hand either over or under the rod, depending upon which direction you wish to bend it. With the rod cradled in the palm of one hand, use your other hand to bend it at the end where the float is. Be careful not to bend the rod sideways. If the float gets too close to the side of the tank, it may stick on the tank. If the float is hung up on the tank, the ball cock cannot work properly. If the ball sticks in the down position, the water does not cut off. If it sticks in the up position, the water does not cut on.

Vertical ball cocks have not been around as long as float-rod ball cocks, but they are very popular. One of the big advantages of a vertical ball cock is that there is no concern about the float sticking to the side of the tank. As float-type ball cocks age, the float rod often becomes loose. When the rod is not screwed into the ball cock securely, the float is apt to come into contact with the side of the tank. Vertical ball cocks remove this risk.

To control the water flow with a vertical ball cock, move the float up and down the shaft of the ball cock. Moving it down produces more water, and moving it up produces less. To move the float, squeeze the thin metal clip, located on a metal rod, next to the float. While squeezing the clip, you can move the float up or down. When you like the placement, release the clip and the float will remain in the selected position.

If a flush valve has a flapper, it may need to be adjusted. If the chain between the flapper and the toilet handle is too short, the flapper will not seal the flush hole. This allows water to run out of the tank and into the bowl. As the water level drops, the ball cock refills the tank. This is an endless cycle that wastes a lot of water. If you notice the ball cock cutting on at odd times or see water running into the bowl long after a flush, investigate the flapper's seal. Moving the chain to the next hole in the handle should improve the seal. If not, adjust the length of the chain where it is connected to the handle. Just make sure the flapper sits firmly on the flush hole.

If the chain between the flapper and the handle is too long, the flapper may become entangled in the chain. This causes the flapper to be held up, allowing water to constantly run down the flush hole. This problem can be corrected by shortening the chain.

Tank balls are sometimes used instead of flappers. If you are dealing with a tank ball, you also have lift wires and a lift-wire guide. The guide is attached to the refill tube in the toilet tank. The lift rods attach to the ball, run through the guide, and attach to the toilet handle. If the alignment of the tank ball is not set properly, water will leak past it, into the toilet bowl.

When you are having problems with the tank ball, inspect the location of the ball on the flush hole. If it is not properly seated, adjust the lift wires or the guide to make a satisfactory seal. The guide can usually be turned with your fingers to make a better alignment. If the lift wires are too short and are holding up the ball, bend them to add length.

If these attempts do not prove fruitful, you may have to replace the tank ball. This is simply a matter of unscrewing the ball from the lift rod and replacing it with a new ball. If a new ball doesn't do the trick, the flush valve may be pitted or defective. On a brass flush valve, you can use sandpaper to remove pits in the brass. By sanding the area where the tank ball sits, you smooth out the rough spots and encourage a better seal. Pits are voids that a tank ball cannot seal. Pits allow water to leak through to the toilet bowl. Unless it is an old toilet, the flush valve should not be pitted.

Tank-to-bowl bolts

As a new toilet is used, the tank-to-bowl bolts may work loose. This is especially true if the tank is not supported by a wall behind it. If leaks develop at the tank-to-bowl bolts, tightening the bolts should stop the leaks. Be careful not to tighten the bolts too much. If you do, the china will break.

If water floods out between the tank and bowl when the toilet is flushed, there is a poor seal at the *flush valve gasket*. This is a gray sponge or black rubber gasket. If the tank-to-bowl bolts are not tight enough to compress this gasket, water will leak each time the toilet is flushed. Normally, tightening the tank-to-bowl bolts stops this type of leak. If not, disassemble the toilet to investigate the problem further.

Possibly the flush valve nut is cracked. Or the threaded portion of the flush valve may have a hole in it. The gasket may have been shifted during installation, rendering it ineffective. A good visual inspection should reveal the cause of the leak. After you make the necessary corrections, reassemble the toilet and test it. If you still have a problem, replace the flush valve.

Kitchen sinks

Kitchen sinks suffer from several problems (Fig. 6.10), one of which is a bad basket strainer. Normally, the only problem with a basket strainer is a poor seal where the drain meets the sink. Removing the drain and reinstalling it with a good seal of putty will stop the leak. If the sink does not hold water, the seal on the basket may be defective. If you suspect the basket is bad, replace it with a new one.

The continuous waste for sinks with multiple bowls may develop leaks after your final test. Usually they will only be small leaks at the slip-nut connections. To stop the leaks, tighten the slip nuts.

Compression fittings under the kitchen sink are the same as those found under the lavatory. It is not unusual for these connections to begin leaking after the system is placed in use. Follow the same methods as given for lavatories to correct deficiencies in the compression fittings under the kitchen sink.

Figure 6.10   Troubleshooting kitchen sinks.

Symptoms	Probable cause
Faucet drips from spout	Bad washers or cartridge Bad faucet seats
Faucet leaks at base of spout	Bad O-ring
Faucet will not shut off	Bad washers or cartridge Bad faucet seats
Poor water pressure	Partially closed valve Clogged aerator Not enough water pressure Blockage in faucet Partially frozen pipe
No water	Closed valve Broken pipe Frozen pipe
Drains slowly	Partial obstruction in drain or trap
Will not drain	Blocked drain or trap
Gurgles as it drains	Partial drainage blockage Partial blockage in vent
Won't hold water	Bad basket strainer Bad putty seal on drain
Spray attachment will not spray	Clogged holes in spray head Kinked spray hose
Spray attachment will not cut off	Bad spray head

The biggest problem with garbage disposals is that they become jammed. If a disposal hums but does not turn, the cutting heads are stuck. To correct this, cut off the power to the disposal. Then place a broom handle in the disposal, and wedge it against the jammed cutting blades. With a little forceful leverage, you should be able to free the blades. Some disposals come with a special wrench, designed to clear the cutting heads. Refer to your owner's instructions to get your disposal back into service.

If the disposal fails to start, try pushing the reset button. Most disposals have a reset button on the bottom of the unit. If pushing the reset button doesn't correct the problem, there may be a blown fuse or a tripped circuit breaker.

This has been a long chapter, but you have learned a lot. Plenty of knowledge is needed when you are working with plumbing fixtures. You may benefit from going back through this chapter later, to pick up on ideas and suggestions that you might have missed.

7

Water Heaters

Most plumbing systems involve the use of water heaters. The heaters are often powered by electricity, but gas, both natural and propane, can be used to generate hot water. Oil-fired water heaters are also available, but not common. Plumbers frequently take water heaters lightly. Many professionals consider installing or replacing a water heater to be a simple job—one that can take less than an hour. Well, the work can be simple, but if it is done improperly, it can have deadly results.

All plumbing codes require permits and inspections for water heater installations and replacements. Even though the code is clear on this, a lot of plumbing contractors never pull permits for jobs where water heaters are being replaced. This is a big mistake that could cost contractors much more than they can afford. I'm not talking about a cash fine or even the loss of a plumbing license. No, I'm talking about someone's death or extreme property damage for which the contractor might be sued.

When you assume the responsibility for installing, replacing, or even working on a water heater, you are getting yourself into a potentially serious situation. When piped improperly, a water heater can become a bomb. It doesn't have to be gas-fired or oil-fired to be dangerous. An electric water heater that is not equipped with the proper safety equipment can blow the roof off a house or scald a person such that major medical attention will be needed. I'm sure you don't want anything like this to happen, let alone have it on your conscience that you were at fault. And certainly, you don't want a court to find you guilty and responsible for hundreds of thousands of dollars in reparations to the person injured or that person's representatives. Water heaters are serious business. Now, let me explain why.

Hot Water

Hot water is provided by most water distribution systems, and some special rules apply to the delivery of hot water. All water heaters must be an approved type. The heaters must be large enough to provide adequate hot water as established by the tables and text found in local code books. They must meet the needs for both daily requirements and peak usage.

Water heaters must have a minimum pressure rating of 125 pounds per square inch (psi), and the pressure rating must be marked permanently on the water heater, in an accessible location. Every water heater must be equipped with a temperature and pressure (T&P) relief valve (Fig. 7.1). Never leave this component out of a system. Without a relief valve, a water heater can blow up. The ratings for T&P valves will be determined by the pressure ratings on the water heaters. However, maximum temperature ratings should not exceed 210°F, and maximum pressure ratings should not exceed 150 psi. These valves are required to monitor the top 6 in of water in the tank.

I've seen water heaters piped so that the sensor tube of a relief valve was barely breaking above the water surface in the heating unit. This is not the right way to install a T&P valve. Don't use nipples and couplings. Screw relief valves into the tapped openings provided for them on water heaters. If you alter the setup, you run the risk of being responsible for severe damages.

Figure 7.1 Temperature and pressure relief valve.

The discharge from a relief valve must be piped downward, to a point of safe discharge. Many water heaters are installed with relief valves that have no discharge tubes. This is dangerous. If the valve releases steam or hot water, someone could be burned badly. It doesn't cost much to install a discharge tube, and code requires one, so make sure you install a tube in compliance with your local code.

The discharge pipe may be connected to the sanitary drainage system indirectly. For example, the discharge pipe could terminate above a floor drain, but it cannot be piped into the drainage system with a closed fitting. Discharge pipes are not allowed to have threaded ends at their point of termination. If they were, someone might install a device that could prevent the proper functioning of the pipe.

The discharge pipes from water heaters are not required to be plumbed into the sanitary drainage system. These pipes may terminate above a floor or at any other point where they do not present a hazard. When the pipes stop above a floor, discharge pipes are commonly held about 6 in above floor level. This is high enough above the floor level to allow a valve to empty itself, without being so high as to pose a threat to bystanders.

When water heaters are installed above habitable space, they may be required to sit in a pan or other container that will protect the living space from flooding, should a problem occur. In such cases, there will be a drain outlet in the retaining pan that should be piped to an appropriate place for the disposal of unwanted water.

Every water heater should be equipped with a drain at its lowest point. Even if the tank is only a hot water storage tank, it should have a drain. Some small water heaters are not equipped with drains. Due to their diminutive size, drains are not a part of the factory product. This is the only exception to the rule about drains.

Water heaters are typically required to meet minimum standards for energy efficiency. There are requirements governing the amount of insulation for water heaters. Probably any water heater you purchase as a new product will conform to these standards.

The controls for adjusting and maintaining the temperature of hot water must be accessible. The temperature controls should function automatically to maintain the desired temperature. When you install a water heater, make sure that the access panels or controls are easy to get at.

Valves should not be installed on the outgoing hot water pipe from a water heater. Neither should valves be installed on the discharge pipes of relief valves. A closed valve in either of these locations could result in an excessive pressure buildup and an explosion.

If the distance between a water heater and a plumbing fixture receiving hot water exceeds 100 ft, steps must be taken to maintain hot water within 100 ft of the fixture.

This might be accomplished by insulating the hot water pipe, but a recirculating system will probably be required. When recirculating pumps are used, they must be equipped with a switch that allows them to be cut off when not in use. Residential systems rarely require such a pump.

Electric water heaters must have a separate switch controlling the power source. If the water heater is fueled by oil or gas, a cutoff valve must be installed on the fuel supply line. All water heaters must have cutoff devices to their fuel or energy source in easy reach of anyone standing near the water heater.

Unions (Figs. 7.2 and 7.3) should be installed in both the hot and cold water pipes near where they attach to a water

Figure 7.2 Closed pipe union.

Figure 7.3 Open pipe union.

Figure 7.4 Gate valve.

Figure 7.5 Stop-and-waste valve.

heater. A full-open valve, such as a gate valve (Fig. 7.4), must be installed on the cold water pipe that feeds into a water heater. Valves that depend on washers to operate, such as stop-and-waste valves (Fig. 7.5) are not acceptable substitutes for full-open valves.

Electric Water Heaters

Electric water heaters can present a number of problems (Fig. 7.6). Their thermostats can go bad, the tanks can leak, the heating elements can burn out, relief valves can pop off, and all sorts of related trouble can come up.

I think it goes without saying, but beware of the electric wires and current involved when working with electric water heaters. As you probably know, there is a lot of voltage running through the wires of a water heater. Once the access cover of an electric water heater is removed, you must be extremely careful not to touch exposed wires and connections.

Relief valves

Relief valves that pop off signal one of three problems: The relief valve is bad, the water heater is building excess pres-

Figure 7.6 Electric water heater.

Symptoms	Probable cause
Relief valve leaks slowly	Bad relief valve
Relief valve blows off periodically	High water temperature High pressure in tank Bad relief valve
No hot water	Electric power off Elements bad Defective thermostat Inlet valve closed
Too little hot water	A bad element Bad thermostat Thermostat needs adjusting
Water too hot	Thermostat needs adjusting Controls defective
Water leaks from tank	Hole in tank Rusted-out fitting in tank

sure, or the heater is building excess temperature. The problem is usually just a defective relief valve. In such cases, replace the relief valve, and monitor it to see that it works properly. If the new valve releases a discharge, look for signs of extreme temperature or pressure in the tank.

The water temperature in the heating tank can be measured with a standard thermometer. Discharge a little water from the relief valve into a container, and test its temperature. If it is too high for the rating of the temperature and pressure relief valve, check the thermostat settings on the water heater. Turn down the heat settings, and test the water again later, after the new temperature settings have been in effect. If the reduction on the thermostat settings does not lower the water temperature in the tank, replacement of the tank is usually the best course of action.

If you suspect the water heater is under too much pressure, test the pressure with a standard pressure gauge. The easiest way to do this is to adapt the gauge to a hose-thread adapter and attach it to the drain at the bottom of the water heater. As long as the drain is not clogged, you can get an accurate pressure reading. You could also adapt the gauge to screw into the relief valve and test the pressure by opening the relief valve.

No hot water

When no hot water is being produced by an electric water heater, only a few things need to be checked. First, check the electric panel to see that the fuse or circuit breaker for the water heater is not blown or tripped. For water heaters that have their own disconnect boxes, check the disconnect lever to see that it is turned on.

Assuming the water heater is receiving adequate electric power, verify that the thermostats are at a reasonable heat setting. It is highly unlikely that anyone would turn them way down, but it is possible.

The most likely cause of this problem is a bad heating element, but a bad thermostat could also be at fault. Check the continuity and voltage of these devices with a meter to determine if they should be replaced.

Limited hot water

When a water heater is producing only a limited amount of hot water, the problem probably lies with the lower heating element. These elements frequently become encrusted with mineral deposits that, in time, reduce the effectiveness of the element. Check the element with your meter, and replace it if necessary.

When a thermostat is set too low, a water heater will not produce adequate hot water. Check the settings on the thermostats, and check the thermostats with your meter.

Water too hot

Sometimes complaints come in that the hot water being produced is too hot. This is normally a simple problem to solve. Turning down the settings on the heater's thermostats will normally correct the situation. However, possibly the thermostat is defective. If it is, replace it.

Complaints of noise

Complaints of noise are sometimes made pertaining to water heaters. If a water heater is installed near habitable space, it is not uncommon for people to notice a rumbling in the heater. This often frightens people into calling a plumber. Do you know what causes the noise being made in the heater? The noise is directly related to sediment that has accumulated in the water heater. Water becomes trapped between layers of the sediment, and when the water temperature reaches a certain point, the water explodes out of the layers. The mini explosions can sound like a cracking or rumbling noise, and they should *not* be ignored. The steam explosions can be controlled by removing sediment from the water heater.

Sediment buildup can be reduced in water heaters by opening the drain valve periodically. Ideally, water heaters should be drained monthly, but few are. If the buildup of sediment is extreme, a cleaning agent can be put into the water heater. The cleaning compound will reduce the scale buildup and the noise.

If you drain water from the heater and don't see any sediment, check the relief valve. The noise may be due to steam buildup. When this is the case, you might need to replace the relief valve.

Rusty water

Rusty water can be a problem with some water heaters. If the hot water from fixtures is rusty, there is a good chance rust has accumulated in the water heater. In moderate cases the rust can be flushed out of the water heater through the drain opening. In extreme cases the entire heater should be replaced.

Leaking tank

When the tank is leaking, plan on replacing the whole unit. It is possible to make temporary repairs for leaking tanks, but they are just that—temporary.

If you need to plug a hole temporarily, you can do it with a toggle bolt and a rubber washer. Drain the water heater to a point below the leak. Drill a hole in the tank that will allow you to insert the toggle bolt. The toggle bolt should have a metal washer against its head and a rubber washer that will come into contact with the tank. Once the toggle penetrates the tank and spreads out, tighten the bolt until the rubber washer is compressed. This will slow down or stop the leak for awhile, but don't expect the repair to last indefinitely. Once a heater starts to develop pinholes, it is time to replace it.

That pretty well covers the range of problems associated with electric water heaters. As long as you know the components you are working with, water heaters are not difficult to troubleshoot. Now let's turn our attention to gas-fired heaters and see how the troubleshooting methods differ.

Gas-Fired Water Heaters

Gas-fired water heaters share some of the same problems (Fig. 7.7) as electric water heaters. However, there are many differences between the two types of heaters. Let's look at the same problems that we studied for electric water heaters and see how your job will differ when you are troubleshooting gas units.

Relief valves

Relief valves on gas heaters can be tested and treated in the same way as those used on electric heaters.

Figure 7.7   Gas water heaters.

Symptoms	Probable cause
Relief valve leaks slowly	Bad relief valve
Relief valve blows off periodically	High water temperature High pressure in tank Bad relief valve
No hot water	Out of gas Pilot light out Bad thermostat Control valve off Gas valve closed
Too little hot water	Bad thermostat Thermostat needs adjusting
Water too hot	Thermostat needs adjusting Controls defective Burner will not shut off
Water leaks from tank	Hole in tank Rusted-out fitting in tank

No hot water

When you are getting no hot water from a gas-fired water heater, first check the pilot light. If the pilot light is not burning, check the gas valve to see that it is turned on. If the gas valve is on and the pilot light is not burning, try to relight the pilot. However, before you do this, make sure there is no accumulation of trapped gas that will explode when you light a flame.

Cut off the gas valve and ventilate the area well. When you are sure it is safe to light the pilot, turn on the gas valve and light the pilot. If the pilot will not light, make sure there is gas coming through the piping. You should be able to hear or smell it.

The thermostat may be turned off or defective. If the thermostat is set in the proper position but won't function, replace it.

The thermocoupling could also be bad. If the pilot lights but continues to go out, replace the thermocoupling.

If none of these methods prove fruitful, check the dip tube. It should be installed in the cold water side of the tank, and it is possible that it was put in the hot water side by mistake.

Limited hot water

When a gas-fired water heater produces only limited hot water, the problem is usually with the gas control. Check the setting to see that it is set high enough to produce a satisfactory supply of hot water. If it is set properly but fails to work, replace it.

The dip tube could be responsible for the production of limited hot water. If it is installed in the hot water side or is broken, either replace it or move it to the cold water side of the tank.

Water too hot

When the water from a gas-fired water heater is too hot, the gas control valve is either bad or set too high. Check the setting, and adjust it to a lower level. If that doesn't solve the problem, replace the valve assembly.

Noise, rusty water, and leaks in tanks

Noise, rusty water, and leaks in tanks can be treated in the same way as for electric water heaters. Refer to the instructions given earlier for electric water heaters.

Gas odors

Gas odors are sometimes noticed around gas-fired water heaters. This can be a dangerous situation. The problem is usually a leak in a fitting, pipe, or piece of tubing. Use soapy water on all places where a leak might occur to find the source of the smell. The soapy water will bubble when it is applied to the site of the leak.

If all the connections, the pipe, and the tubing check out okay, inspect the venting of the water heater. Inadequate or improper venting could cause gas smells to be trapped around the heater.

Oil-Fired Water Heaters

Oil-fired water heaters are not common in most areas, but they do exist. These heaters are typically large. They are sometimes used for residential applications, but most of the ones that I've worked with have been used in commercial properties. This type of heater requires a different approach from that used for electric or gas heaters.

In the jurisdictions where I have worked, plumbers are usually licensed to work with gas. I know this is not the case in all places, but it has been true in my work areas. Since I'm both a master plumber and a master gasfitter, I'm permitted to work with both electric and gas systems completely. This is not true of oil-fired units.

Since oil-fired water heaters are fueled by oil, an oil burner repair license is usually needed to work on the fuel portion of the system. Basic troubleshooting of an oil-fired unit is very similar to the procedures used with a gas-fired water heater, so you can refer to the instructions in the section on gas heaters. However, if there is a problem with the unit firing or any other aspect of the oil setup, you will probably have to defer to a licensed oil technician. Keep this in mind, and don't offer your services for work that is not covered under your license or insurance. This same advice might apply to gas-fired units if you are not licensed to work with gas.

Water heaters are limited in the type of defects that they may suffer from. This makes troubleshooting them fairly easy. The information in this chapter should steer you in the right direction. As with any troubleshooting situation, it is helpful to have a manual on hand for the particular

unit you are working with, although this is rarely necessary with water heaters. As a final word of advice, don't
install or replace water heaters without a permit. The
reward is not worth the risk.

8

Well Systems

Plumbers in rural locations work with well systems regularly. They install them, and they service them. City-based plumbers have little experience with well systems. Even plumbers in the country don't always have a lot of experience with well systems. In some regions, well drillers control most of the pump business. But if you work in an area where water wells and pumps are used, you should be prepared for the service calls you may get.

There is good money to be made in well systems. Whether you are installing new systems or fixing problems with existing pumps, the pay can be lucrative. How much do you have to know? It depends on what your plans are. It takes less knowledge to install new well systems than it does to troubleshoot and repair them.

Most pump makers offer booklets to professional plumbers (through plumbing suppliers) that offer troubleshooting tips and techniques. These publications are invaluable when you are in the field with a pump system that you are not familiar with. Ask your supplier for booklets on all the brands of pumps used in your area. You will probably have to visit several suppliers to obtain literature on all the various types of pumps. If suppliers can't help you, contact the pump makers directly. They will be happy to send you all sorts of information on their products. When you combine specific manufacturer information with the data in this chapter, you will be a formidable force in the installation and repair of well systems.

Shallow-Well Jet Pumps

Shallow-well pumps are used with wells where the lowest water level will not be more than 25 ft below the pump. Many factors influence the type and size of pump that an installation will require. The first consideration is the height to which water will be pumped. If a pump has to pump water higher than 25 ft, a shallow-well pump is not a viable choice. Shallow-well pumps are not intended to lift water more than 25 ft. This limitation is due to the way that a shallow-well pump works.

These pumps work on a suction principle. The pump sucks the water up the pipe and into the home. With a perfect vacuum at sea level, a shallow-well pump may be able to lift water to 30 ft. This maximum lift is not recommended and is rarely achieved. If the pump has to lift water higher than 25 ft, investigate other types of pumps. When conditions allow the use of a shallow-well jet pump, this section tells you how to install it. Talk to your pump dealer or refer to the manufacturer's recommendations for proper sizing of the pump.

A suction pump

When you install a suction pump, the single pipe from the well to the pump usually has a diameter of $1\frac{1}{4}$ in. A standard well pipe material is polyethylene, rated for 160 psi. Make sure that the suction pipe is not coiled or in any condition that may cause it to trap air as it is being installed. If the pipe holds an air pocket, priming the pump can be quite difficult. In most cases a foot valve is installed on the end of the pipe that is submerged in the well.

Screw a male insert adapter into the foot valve. Place

two stainless-steel clamps over the well pipe, and slide the insert fitting into the pipe. Tighten the clamps to secure the pipe to the insert fitting. When you lower the pipe and foot valve into the well, don't let the foot valve sit on the bottom of the well. If the suction pipe is too close to the bottom of the well, it may suck sand, sediment, or gravel into the foot valve. If this happens, the pipe cannot pull water from the well.

When the pipe reaches the upper portion of the well, it usually takes a 90° turn to exit the well casing. This turn is made with an insert-type elbow. Always use two clamps to hold the pipe to its fittings. When the pipe leaves the well, it should be buried underground. Run the well pipe through a sleeve so that the well casing will not chafe the pipe during use and wear a hole in it. The pipe must be deep enough that it will not freeze in the winter. This depth will vary from state to state. Your local plumbing inspector can tell you to what depth you must bury the water supply pipe.

When you place the pipe in the trench, be careful not to lay it on sharp rocks or other objects that might wear a hole in it. Backfill the trench with clean fill dirt. If you dump rocks and cluttered fill on the pipe, it can be crimped or cut. When you bring the pipe into the home, run it through a sleeve as it comes through or under the foundation. The sleeve should be two pipe sizes larger than the water supply pipe.

Once the pipe is inside the home, you may have to convert it to copper, CPVC, PB, or one of the other approved materials. When PE pipe is used as a water service, it must be converted to some other type of pipe within 5 ft of entry into the home, assuming that the house will have both hot and cold water. If you convert the pipe, the conversion will typically be done with a male insert adapter. The water supply pipe should run directly to the pump. The foot valve acts as a strainer and as a check valve. When there is a foot valve in the well, there is no need for a check valve at the pump.

The incoming pipe will attach to the pump at the inlet opening with a male adapter. At the outlet opening, install a short nipple and a tee fitting. At the top of the tee, install reducing bushings and a pressure gauge. From the center outlet of the tee, the pipe will run to another tee fitting. A gate valve should be installed in this section of pipe, near the pump. At the next tee, the center outlet is piped to a pressure tank. From the end outlet of the tee, the pipe will run to yet another tee fitting. At this tee, the center outlet becomes the main cold water pipe for the house. Another gate valve should be installed in the pipe feeding the water distribution system. On the end outlet of the tee, install a pressure relief valve. All these tee fittings should be in a close proximity to the pressure tank.

The pump is equipped with a control box that requires electric wiring. This job should be done by a licensed electrician. If you are an electrician, you know how to do the job. If you are not an electrician, do *not* attempt to wire the controls.

Priming the pump

The pump has a removable plug in the top of it, to allow the priming of the pump. Remove the plug, and pour water into the priming hole. Continue this process until the water is standing in the pump and visible at the hole. Apply pipe dope to the plug, and screw it back into the pump. When you turn on the pump, you should have water pressure. If you don't, continue the priming process until the pump is pumping water. This can be a time-consuming process; don't give up.

Setting the water pressure

Once the pump is pumping water and filling the pressure tank, setting the water pressure is a priority. When the tank is filled, the pressure gauge should read between 40 and 60 psi. The pump's controls are preset at cut-on and cutoff intervals. These settings regulate when the pump cuts on and off. Typically, a pump cuts on when the tank pressure drops below 20 psi. The pump cuts off when the tank pressure reaches 40 psi.

If your customer prefers a higher water pressure, the pressure switch can be altered to deliver higher pressure. You might have the controls set to cut on at 40 psi and off at 60 psi. These settings are adjusted inside the pressure switch, around electric wires. There is possible danger of electrocution when you are making these adjustments. To avoid problems, cut off the power to the pressure switch while you are making the adjustments.

The adjustments are made by turning a nut that sits on top of a spring in the control box. You will see a coiled spring, compressed with a retaining nut. By moving this nut up and down the threaded shaft, you can alter the cut-on and cutoff intervals. Refer to the manufacturer's recommendations for precise settings.

Deep-Well Jet Pumps

When the water level is more than 25 ft below the pump, you have to use either a deep-well jet pump or a submersible pump. In today's plumbing applications, submersible pumps are normally used in deep wells. However, deep-well jet pumps will get the job done.

Deep-well jet pumps resemble shallow-well pumps. They are installed above ground and are piped in a similar manner to shallow-well pumps. The noticeable difference is the number of pipes going into the well. A shallow-well pump has only one pipe. Deep-well jet pumps have two pipes. The operating principles of the two types of pumps differ. Shallow-well pumps suck water up from the well. Deep-well jet pumps push water down one pipe and suck water up the other; this is why there are two pipes on deep-well jet pumps.

The only major difference between a shallow-well pump and a deep-well pump is the number of pipes used in the installation and a pressure control valve. Deep-well pumps still use a foot valve. A jet body fitting is submerged in the well and attached to both pipes and the foot valve. The pressure pipe connects to the jet assembly first. The foot valve hangs below the pressure pipe. There is a molded fitting on the jet body for the suction line to connect to. With this jet body, both pipes are allowed to connect in a natural and efficient manner.

Deep-well jet pumps exert pressure *down* the pressure pipe; with water pushed through the jet assembly, it makes it possible for the suction pipe to pull water *up* from the deep well. From the suction pipe, water is brought into the pump and is distributed to the potable water system. At the head of a deep-well jet pump, you will see two openings for the pipes to connect to. The larger opening is for the suction pipe, and the smaller opening is for the pressure pipe. The suction pipe usually has a diameter of $1\frac{1}{4}$ in. The pressure pipe typically has a diameter of 1 in.

The piping from the pump to the pressure tank needs a pressure control valve installed in it. This valve ensures a minimum operating pressure for the jet assembly. Shallow-well pumps are not required to have a pressure control valve. Once the pressure control valve is installed, the remainder of the piping is done in the same manner as for a shallow-well pump.

Submersible Pumps

Submersible pumps are very different from jet pumps. Jet pumps are installed outside the well. Submersible pumps are installed in the well, submerged in the water. Jet pumps use suction pipes. Submersible pumps have only one pipe, and they push water up the pipe, from the well. Jet pumps use a foot valve; submersible pumps don't. Submersible pumps are much more efficient than jet pumps; they are also easier to install. Under the same conditions, a 0.5-horsepower (hp) submersible pump can produce nearly 300 gal more water than a 0.5-hp jet pump. With so many advantages, it is almost foolish to use a jet pump, when you could use a submersible pump. The one exception occurs when you are installing a pump for a shallow well. Then a jet pump makes the most sense.

Installing a submersible pump requires different techniques from those used with a jet pump. Since submersible pumps are installed in the well, electric wires must run down the well to the pump. Before you install a submersible pump, consult a licensed electrician about the wiring needs of the pump. Some plumbers do this wiring themselves, but many jurisdictions require the connections to be made by a licensed electrician.

You will need a hole in the well casing to install a pitless adapter. The pitless adapter provides a watertight seal in the well casing for the well pipe to feed the water service. When you buy the pitless adapter, it should be packaged with instructions on what size hole is needed in the well casing.

Cut a hole in the well casing with a cutting torch or a hole saw. The pitless adapter attaches to the well casing and seals the hole. On the inside of the well casing, there is a tee fitting on the pitless adapter. This is where the well pipe is attached. This tee fitting is designed to allow you to make all pump and pipe connections above ground. After all the connections are made, lower the pump and pipe into the well, and the tee fitting slides into a groove on the pitless adapter.

To make up the pump and pipe connections, you need to know how deep the well is. The well driller should provide you with the depth and rate of recovery for the well. Once you know the depth, cut a piece of plastic well pipe to the desired length. The pump should hang at least 10 ft above the bottom of the well and at least 10 ft below the lowest expected water level.

Apply pipe dope to a male-insert adapter, and screw it into the pump. This fitting is normally made of brass. Slide a torque arrester over the end of the pipe. Next, slide two stainless-steel clamps over the pipe. Place the pipe over the insert adapter, and tighten the two clamps. Compress the torque arrester to a size slightly smaller than the well casing, and secure it to the pipe. The torque arrester absorbs thrust and vibrations from the pump and helps to keep the pump centered in the casing.

Slide the torque stops down the pipe from the end opposite the pump. Space the torque stops at routine intervals along the pipe to prevent the pipe and wires from scraping against the casing during operation. Secure the electric wiring to the well pipe at regular intervals to eliminate slack in the wires. Apply pipe dope to a brass male-insert adapter, and screw it into the bottom of the tee fitting for the pitless adapter. Slide two stainless-steel clamps over the open end of the pipe, and push the pipe onto the insert adapter. After tightening the clamps, you are ready to lower the pump into the well.

Before you lower the pump, it is a good idea to tie a safety rope onto the pump. After the pump is installed, this rope will be tied at the top of the casing to prevent the pump from being lost in the well, if the pipe becomes disconnected from the pump. Next, screw a piece of pipe or an adapter into the top of the tee fitting for the pitless adapter. Most plumbers use a rigid piece of steel pipe for this purpose.

Once you have a pipe extending up from the top of the tee fitting, lower the whole assembly into the well casing. This job is easier if you have someone to help you. Be careful not to scrape the electric wires on the well casing as you lower the pump. If the insulation on the wires is damaged, the pump may not work. Holding the assembly by the pipe extending from the top of the pitless tee, guide the pitless adapter into the groove of the adapter in the well casing. When the adapter is in the groove, push down to seat it in the mounting bracket. This concludes the well part of the installation.

Attach the water service pipe to the pitless adapter on the outside of the casing. You can do this with a male-insert adapter. Run the pipe to the house in the same way as described for jet pumps. Once the pipe is inside the house, install a union on the water pipe. The next fitting should be a gate valve, followed by a check valve. From the check valve, the pipe should run to a tank tee.

The tank tee is a device that screws into the pressure tank and allows the installation of all related parts. The switch box, pressure gauge, and boiler drain can all be installed on the tank tee. When the pipe comes to the tank tee, the water is dispersed to the pressure tank, the drain valve, and the water main. Where the water main leaves the tank tee, you should install a tee to accommodate a pressure relief valve. After this tee, you can install a gate valve and continue piping to the water distribution system. All that is left is to test the system; you do not have to prime a submersible pump.

Pump Problems

Pump problems are not uncommon. If you offer your services as a plumber who troubleshoots and repairs pumps, you have to be prepared for whatever you might find. I've already told you that pump makers often offer troubleshooting booklets to professionals. Take advantage of this. Many problems are the same from pump to pump, regardless of brand, but it is always best to have information specific to the particular type of pump that you are working with.

I can't describe every detail for all the various types of pumps being used. But I can give you some solid troubleshooting tips that will often apply to all types of pumps. Servicing pumps is sometimes a matter of trial-and-error testing, but if you have the right guidelines to follow, you can shorten the time you spend trying to identify a problem. Let's start with jet pumps.

Troubleshooting Jet Pumps

When a jet pump will not run

Circuit breakers and fuses.   When a pump won't run (Fig. 8.1), there are five likely causes of the malfunction. First

Figure 8.1   Jet potable-water pumps.

Symptoms	Probable cause
Won't start	No electric power
	Wrong voltage
	Bad pressure switch
	Bad electrical connection
	Bad motor
	Motor contacts open
	Motor shaft seized
Runs, but produces no water	Needs to be primed
	Foot valve is above water level in well
	Strainer clogged
	Suction leak
Starts and stops too often	Leak in piping
	Bad pressure switch
	Bad air control valve
	Waterlogged pressure tank
	Leak in pressure tank
Low water pressure in pressure tank	Strainer on foot valve partially blocked
	Leak in piping
	Bad air charger
	Worn impeller hub
	Lift demand too much for pump
Pump does not cut off when working pressure is obtained	Pressure switch bad
	Pressure switch needs adjusting
	Blockage in piping

check the fuse or circuit breaker that controls the pump's electric circuit. If the fuse is blown or the breaker is tripped, the pump cannot run. When you check the electric panel and find these conditions, replace the blown fuse or reset the tripped breaker. In future references to fuses or circuit breakers, I will only refer to circuit breakers. If a home has fuses, when I say to reset the circuit breaker, replace the blown fuse.

Damaged or loose wiring.   If the circuit breaker is not tripped, inspect all the electric wiring that affects the pump. A broken or loose wire may be preventing the pump from running. Remember, you are dealing with electricity and must use appropriate caution to avoid electric shocks.

Pressure switches.   Pressure switches can become defective. It is also possible that the switch is out of proper adjustment. To correct this problem, adjust or replace the pressure switch. Remove the cover of a pressure switch; you will see two nuts sitting on top of springs. The nut on the short spring is set at the factory and should not need any adjustment. However, if you want a pump to cut off at a higher pressure, turn this nut clockwise. To have the pump cut off at a lower pressure, turn the nut counterclockwise. The nut on the taller spring controls the cut-on/cutoff cycle of your pump. To make the pump cut on at a higher pressure, turn the nut clockwise. If you want the pump to cut on at a lower pressure, turn the nut counterclockwise.

Stopped-up nipples.   If the tubing or nipples on the pressure switch become restricted, the pump may not run. If you suspect that this is the problem, take apart the pipe and fittings and clean or replace them. In a new installation, this problem is unlikely, but it is possible.

Overloaded motors.   If the motor becomes overloaded, protection contacts will remain open. This problem will solve itself. The contacts will close automatically after a short time.

Seized pumps.   If a pump becomes mechanically bound, it will not run. To correct this problem, first remove the end cap from the pump. Now you should be able to turn the motor shaft with your hands. If the shaft rotates freely, reassemble the pump and test it. The problem should be corrected.

Voltage problems.   If a pump is not receiving proper voltage, it will not run. This problem calls for direct contact with the electrical system. Because of the danger involved in working with electricity, I cannot help you with this aspect of troubleshooting. If you have eliminated all the nonelectrical problems, call in a professional electrician to fix the pump.

Defective motors.   Obviously, defective motors can prevent pumps from running. To determine if a pump's motor is bad, you must do extensive electrical troubleshooting. Check the ground, capacitor, switch, overload protector, and winding continuity. If you are experienced in working with electrical systems, you already know how to do this. If you are not, I am afraid to give you instructions on such a dangerous job. One wrong move could deliver a potent electric shock. Again, if the problem is electrical, call in a professional electrician.

Pump runs, but produces no water

Loss of the pump's prime.   If a pump loses its prime, the pump will run without producing water. With a shallow-well pump, remove the priming plug and pour water into the pump. When the water level stands static at the opening, apply pipe dope to the priming plug and replace it. Start the pump, and you should get water pressure. With a deep-well jet pump, trying to prime the pump through the priming hole is a losing proposition.

With a two-pipe system, disconnect the pipes from the pump. Pour water down each pipe until they are both full of water. When water has filled both pipes and is holding its level, reconnect the pipes to the pump. Prime the pump through the priming hole. Start the pump, and you should get water. If you don't, the problem may be in the pressure control valve.

Pressure control valves.   In a two-pipe system, the pressure control valve may need to be adjusted when a pump runs without producing water. When the pressure control valve is set too high, the air volume control may not function. When the pressure control valve is set too low, the pump may not cut off. To reduce the water pressure, turn the adjusting screw on the pressure-reducing valve counterclockwise. To increase pressure, turn the adjusting screw clockwise.

Lack of water. If a well pipe is not below the water level, the pump cannot produce water. With shallow wells, it is not uncommon for the water levels to drop during hot, dry months. If a pump is running but is not producing water, check the water level in the well. You can do this with a roll of string and a weight. Tie the weight on the end of the string, and lower it into the well. When the weight hits bottom, withdraw the string. The end of the string should be wet. Measure the distance along the wet string. This will tell you how many feet of water are in the well. This information, along with the records of how long the well pipe is, will tell you if the end of the pipe is submerged in water.

Clogged foot valves. If a foot valve is too close to the bottom of the well, it may become clogged with mud, sand, or sediment. It is not uncommon for shallow wells to fill in with sediment. What starts out as a 30-ft well may become a 25-ft well as the earth settles and shifts. Even if a foot valve was originally set at an appropriate height, it may now be hanging too low in the well. Sometimes, shaking the well pipe will clear the foot valve of its restriction. In other cases, you have to pull the pipe out of the well and clean or replace the foot valve. If the foot valve is pulling up sediment from the bottom of the well, shorten the well pipe.

Malfunctioning valves. Foot valves act as a check valve. If the foot valve or check valve is stuck in the closed position, you will not get any water from the pump. Pull the well line and inspect the check valve or foot valve. If the valve is stuck, replace it.

Defective air volume control. It is possible that the diaphragm in an air volume control has a hole in it. By disconnecting the tubing and plugging the connection in the pump, you can determine whether there is a hole in the diaphragm of the air volume control. If plugging the hole corrects the problem, you must replace the air volume control.

Failing jet assemblies. If the jet assembly fails, the pump will not produce water. Inspect the jet assembly for possible obstructions. If you find an obstruction, try to clear it. If you cannot clean the assembly, replace it.

Suction leaks. A common ailment of shallow-well systems is a leak in the suction pipe. To check for this problem, you must pressurize the system and inspect it for leaks. If you find a leak, plug it and you will have water again.

Pump will not cut off

If a pump builds pressure, but does not cut off, there may be a bad pressure switch. Before you replace the switch, there are a few things you should check. Check all the tubing, nipples, and fittings associated with the switch. If they are obstructed, clear them and test the pump. If the problem persists, check that the pressure switch is set properly. Do this by checking the cut-on and cutoff controls discussed earlier. If all else fails, replace the pressure switch.

Pump doesn't get enough pressure

A lack of water pressure can be caused by several problems. If a pump is trying to lift water too high, the pump may not be able to handle the job. This requires checking the rating of the pump to be sure it is capable of performing under the given conditions. If there are leaks in the piping, the pump will not be able to build adequate pressure. Check all piping to confirm that there are no leaks. If the jet or the foot valve is partially obstructed, the water pressure will suffer. Check these items to be sure they are free to operate. If they are blocked, clear the obstructions and test the pump.

When an air volume control is on the blink, you may not be able to build suitable pressure. Test the control as described above, and replace it if necessary. On older pumps, the impeller hub or guide vane bore may be worn.

If this is the case, call a pump expert to verify your diagnosis. If either of these parts is worn, it must be replaced.

Pump cycles too often

Leaks in piping can cause a pump to cut on and off more frequently than it should. When the system is pressurized, check all piping for leaks. If any are found, repair them. If the pressure switch is not working correctly, the pump may run randomly. Follow the directions given earlier to troubleshoot the pressure switch. If necessary, replace the switch. The problem could be in the suction lift of the system. If water floods the pump through the suction line, you must control the water flow with a partially closed valve. Also there may not be enough vacuum on the suction line. During troubleshooting, check out the air volume control. Perhaps this valve is defective. The instructions given earlier explain how to check the air volume control.

Pressure tanks.   The most likely cause of pumps that cycle too often is a faulty pressure tank. If the tank has a leak in it, it cannot hold pressure. In old tanks, this is a common problem. If you find a leak in the tank, replace the tank. If the air pressure in the pressure tank is not right, the pump will cut on frequently. The pump may cut on every time water to a faucet is turned on. This condition should not be ignored; it could burn out the motor of the pump.

Modern captive air tanks come with a factory precharge of air. These tanks use a diaphragm to control the air volume. Older tanks do not have these diaphragms. Older tanks commonly become waterlogged. When the tank is waterlogged, it does not contain enough air and does contain too much water.

Air pressure.   To check the air pressure in a tank, cut off the electric power to the pump. Drain the tank until the pressure gauge reads zero. A rule of thumb for proper air charge is that the air pressure should be 2 psi less than the cut-on pressure of the pressure switch. For example, if the pressure switch is set to cut on at 40 psi, the air in the tank should be set at 38 psi. You can check the air pressure by putting a tire gauge on the air valve of the tank.

If the air pressure is too low, pump air into the tank. You can do this with a bicycle pump or an air compressor. Monitor the pressure as you fill the tank with air. When the air pressure is at the proper setting, turn the electric power back on to the pump. When the water floods the tank, the pressure gauge should have a normal reading. The system should work properly if the air pressure remains at the proper setting.

If the pressure tank is old, it may have a small leak that allows the air to escape. It is possible that a new tank is defective. If this procedure does not work and keep the system working, consider replacing the pressure tank.

Submersible Pumps

Pump won't start

Circuit breakers.   The first item to check when a submersible pump won't start (Fig. 8.2) is the circuit breaker for the pump. If the breaker is tripped, the pump cannot run. If you check the electric panel and find the breaker tripped, reset the tripped breaker.

Voltage problems.   If the circuit breaker is not tripped, the voltage to the pump must be checked. This inspection can be done quickly with a voltage meter.

Pump starts but does not run

Submersible pumps that start but do not run should be checked in the following manner:

Fuses.   Check all fuses or circuit breakers to see that they are not blown or tripped. If improper fuses are installed, the pump may start but not run.

Voltage.   Incorrect voltage can allow the pump to start, but not allow it to continue running.

Figure 8.2 Submersible potable water pumps.

Symptom	Probable cause
Won't start	No electric power
	Wrong voltage
	Bad pressure switch
	Bad electrical connection
Starts, but shuts off fast	Circuit breaker or fuse inadequate
	Wrong voltage
	Bad control box
	Bad electrical connections
	Bad pressure switch
	Pipe blockage
	Pump seized
	Control box too hot
Runs, but does not produce water or produces only a small quantity	Check valve stuck in closed position
	Check valve installed backward
	Bad electric wiring
	Wrong voltage
	Pump sitting above water in well
	Leak in piping
	Bad pump or motor
	Broken pump shaft
	Clogged strainer
	Jammed impeller
Low water pressure in pressure tank	Pressure switch needs adjusting
	Bad pump
	Leak in piping
	Wrong voltage
Pump runs too often	Check valve stuck open
	Pressure tank waterlogged and needs air injected
	Pressure switch needs adjusting
	Leak in piping
	Wrong-size pressure tank

Seized pump. If a pump has seized with sand or gravel, pull the pump and clean it. When the pump has seized, a high amperage reading is found from tests with a meter.

Damaged wiring. If the electric wiring to the pump is damaged, the pump may not continue to run. This condition can be tested with a meter by reading the resistance. The insu-

lation on the wires may have been damaged when the
pump was lowered into the well. A spliced connection may
have become open or short-circuited. When the problem is
located, the wiring must be repaired or replaced.

Pressure switch.   Your problem may be caused by a defective pressure switch. Check the points on the pressure
switch, and replace it if the points are bad.

Pipe obstructions.   If the tubing or nipples for the pressure
switch are clogged, the pump may not run. Check for
obstructions and clear the tubing or nipples if necessary.

Loose wires.   It is possible that you are a victim of loose
wires. Check the wiring at the control box and the motor. If
the wires are loose, the pump may start, then stop.

Control box.   If the pump has never run properly, it may
have been fitted with the wrong control box. Inspect the
box and confirm that it is the correct one for the pump.
When the control box is exposed to high temperatures, customers may experience trouble with pumps. If a control box
is located in an area where the temperature exceeds 120°F,
the high temperature may be causing the problem.

Pump runs but gives no water

Clogs.   If the protective screening on the pump is obstructed, water cannot enter the pump. This will cause the pump
to run without providing water. When impellers are
blocked by obstructions, you will experience the same problem. Pull the pump and check the strainer and impellers
for clogs.

Bad motor.   If the pump motor is badly worn, it may not
produce water, even if it runs. You will have to pull the
pump to inspect the motor. If it is worn, replace it.

Electrical problems.   Improper voltage, incorrect connections, or loose connections could be the cause of your problem. You might discover these problems with an electric

meter. If you are not comfortable working with electric wires, call an electrician to help you.

Leaks. If there are leaks in the piping, water may not be able to get to the house. Connect an air compressor to the pipe, and pump it up with air to test for leaks. Inspect all piping for leaks, and repair or replace the pipe as needed.

Check valves. Sometimes a check valve is installed backward. If there are problems with a new installation, check that the check valve was properly installed. It is also possible that the check valve has stuck in the closed position. Do a visual inspection, and replace or reinstall the check valve if necessary.

No water. Normally, drilled wells maintain a good water level, but at times the water level drops below the pump. If the pump is not submerged, it cannot pump water. Check the water level in the well with a string and a weight. If the water is low, you may be able to lower the pump; but don't hang the pump too close to the bottom of the well.

When a deep well is drilled, the installer should provide a recovery rate for the well. The recovery rate is stated in gallons per minute (gpm). The minimum recovery rate normally accepted is 3 gpm, and 5 gpm is a better rate. If a pump is pumping faster than the well can recover, the well may run out of water. Pumps are rated according to how many gallons per minute they pump. Compare the pump's rating to the recovery rate of the well, and you may find the cause of the trouble.

Low water pressure

If there are leaks in the piping, a pump will not be able to build adequate pressure. Check all piping to confirm that no leaks are present. Check the setting on the pressure switch. An improper setting will affect water pressure. If the pump is worn or the voltage to the pump is incorrect, you may not have strong water pressure.

Pump cycles too often

If a check valve is stuck in the open position, a pump will have to work overtime. Inspect the check valve to be sure it is operating freely. If the valve is bad, replace it. Leaks in the piping may cause a pump to cut on and off frequently. Check all piping for leaks. If any are found, repair them. If the pressure switch is not working correctly, the pump may cycle too often. Follow the directions given earlier to troubleshoot the pressure switch. If necessary, replace the switch.

The most likely cause of this problem is a faulty pressure tank. If the tank has a leak in it, it cannot hold pressure. In old tanks, this is a common problem. If you find a leak in the tank, replace the tank. If the air pressure in the pressure tank is not right, the pump will cut on frequently. The pump may cut on every time water to a faucet is turned on. This condition should not be ignored; it could burn out the pump motor.

Modern captive air tanks come with a factory precharge of air. These tanks use a diaphragm to control the air volume. Older tanks do not have these diaphragms. Older tanks commonly become waterlogged. When the tank is waterlogged, it does not contain enough air and does contain too much water.

To check a pressure tank's air pressure, cut off the electric power to the pump. Drain the tank until the pressure gauge is at zero. A rule of thumb for the proper air charge is that the air pressure should be 2 psi less than the cut-on pressure of the pressure switch. For example, if a pressure switch is set to cut on at 40 psi, air in the tank should be set at 38 psi. Check the air pressure by putting a tire gauge on the air valve of the tank.

If the air pressure is too low, pump air into the tank. You can use a bicycle pump or an air compressor. Monitor the pressure as you fill the tank with air. When the air pressure is at the proper setting, turn the electric power back on to the pump. When the water floods the tank, the pressure gauge should have a normal reading. The system should work properly if the air pressure remains at the proper setting. The pressure tank may have a small leak in it, allowing air to escape. If the tank cannot hold air, it cannot build pressure.

Well, there you have it. With the help of this chapter, you should be able to solve most problems associated with pump systems. In the next chapter we talk about septic systems. After all, if you are working with well pumps, you are probably working in an area where septic systems are common.

9

Septic Systems

Septic systems are common in rural housing locations. Many people who live outside the parameters of municipal sewers depend on septic systems to solve their sewage disposal problems. Plumbers who work in areas where private waste disposal systems are common often come into contact with problems associated with septic systems. Ironically, plumbers are rarely the right people to call for septic problems, but they are often the first people that homeowners think of when experiencing septic trouble.

One reason plumbers are called so frequently for septic problems is that the trouble appears to be a stopped-up drain. When a septic system is filled beyond capacity, backups occur in houses. Most homeowners call plumbers when this happens. Smart plumbers check the septic systems first to find out whether they are at fault.

Backups in homes are not the only reason why plumbers need to know a little something about septic systems. Customers frequently have questions about their plumbing systems that can be influenced by a septic system. For example, is it all right to install a garbage disposal in a home that is served by a septic system? Some people think it is, and others believe it isn't. The answer to this question may not be left to a plumber's personal opinion. Many local plumbing codes prohibit the installation of food grinders in homes where a septic system will receive the discharge.

Given all the questions and concerns that customers might raise with their plumbers, I think it is wise for plumbers to acquire general knowledge of septic systems. This chapter will help you achieve this goal. With that said, let me show you what is involved with septic systems.

Simple Septic Systems

Simple septic systems consist of a tank, some pipe, and some gravel. These systems are common, but they don't work well in all types of ground. Since most plumbers are not septic installers, I will not bore you will all the sticky details of putting a pipe-and-gravel system into operation. However, I will give a general overview of the system so that you can talk intelligently with customers.

The Components

Let's talk about the basic components of a pipe-and-gravel septic system. Starting near the foundation of a building, there is a sewer. The sewer pipe should be made of solid pipe, not perforated pipe. I know this seems obvious, but I did find a house about 3 years ago where perforated drain-field pipe was used in the sewer. It was quite a mess. Most jobs today involve the use of schedule 40 plastic pipe for the sewer. Cast-iron pipe can be used, but plastic is the most common and is certainly acceptable.

The sewer pipe runs to the septic tank (Fig. 9.1). Septic tanks can be made of many types of materials, but most are constructed of concrete. It is possible to build a septic tank on-site, but every contractor I've ever known has bought precast tanks. An average tank holds about 1000 gal. The connection between the sewer and the septic tank should be watertight.

The discharge pipe from the septic tank should be made of solid pipe, just as the sewer pipe is. This pipe runs from the septic tank to a distribution box, which is also normally

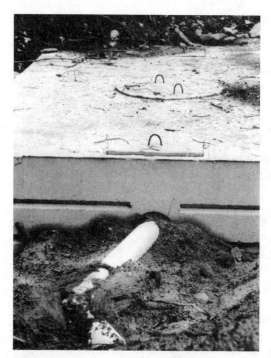

Figure 9.1 Common septic tank installation.

made of concrete. Once the discharge pipe reaches the distribution box, the type of materials used changes.

The drain field is constructed according to an approved septic design. In basic terms, the excavated area for the septic bed is lined with crushed stone. Perforated plastic pipe is installed in rows. The distance between the drain pipes and the number of drain pipes are controlled by the septic system design. All the drain-field pipes connect to the distribution box. The septic field is then covered with material specified in the septic system design.

As you can see, the list of materials is not long. Some schedule 40 plastic pipe, a septic tank, a distribution box, some crushed stone, and some perforated plastic pipe are the main ingredients. This is the primary reason why the cost of a pipe-and-gravel system is so low compared to that of other types of systems.

Types of tanks

There are many types of septic tanks in use today. Precast concrete tanks are by far the most common. Since they are not the only type of septic tank available, let's discuss some of the material options.

Precast concrete is the most popular type of septic tank. When this type is installed properly and is not abused, it can last almost indefinitely. However, heavy vehicular traffic running over the tank can damage it, so this should be avoided.

Metal septic tanks were once prolific. There are still lots of them in use, but new installations rarely use a metal tank. The reason is simple: Metal tends to rust, and that's not good for a septic tank. Some metal tanks are said to have given 20 years of good service. This may be true, but there are no guarantees that a metal tank will last even 10 years. In all my years as a contractor, I've never seen a metal septic tank installed. I've dug up old ones, but I've never seen a new one go in the ground.

I don't have any personal experience with fiberglass septic tanks, but I can see some advantages to them. Their light weight is one nice benefit for any installer. Durability is another strong point in the favor of fiberglass tanks. However, I'm not sure how the tanks perform under the stress of being buried. I assume that their performance is good, but I have no first-hand experience.

Wood seems like a strange material to use for the construction of a septic tank, but I've read about its being used. The wood of choice is redwood. I guess if you can make hot tubs and spas out of it, you can make a septic tank out of it. However, I would not want to warranty a septic tank made of wood.

Brick and block have also been used to form septic tanks.

When these materials are employed, some type of parging and waterproofing must be done on the interior of the vessel. Personally, I would not feel very comfortable with this type of setup. This is, again, material that I have never worked with in the creation of a septic tank, so I can't offer much in the way of case histories.

Chamber Systems

Chamber septic systems are used most often when the percolation rate on ground is low. Soil with a rapid absorption rate can support a standard, pipe-and-gravel septic system. Clay and other types of soil may not. When bedrock is close to the ground surface, chambers are often used.

What is a chamber system? A chamber system is installed in much the same manner as a pipe-and-gravel system, except for the use of chambers. The chambers might be made of concrete or plastic. Concrete chambers are naturally more expensive to install. Plastic chambers are shipped in halves and are put together in the field. Since plastic is a very durable material and it's relatively cheap, plastic chambers are more popular than concrete chambers.

When a chamber system is called for, typically many chambers are involved. These chambers are installed in the leach field, between sections of pipe. As effluent is released from a septic tank, it is sent into the chambers. The chambers collect and hold the effluent for a time. Gradually, the liquid is released into the leach field and is absorbed by the earth. The primary role of the chambers is to retard the distribution rate of the effluent.

Building a chamber system allows you to take advantage of land that would not be buildable with a standard pipe-and-gravel system. On this basis, chamber systems are good. However, when you look at the price tag of a chamber system, you may need a few moments to catch your breath. I've seen a number of quotes for these systems that pushed the $10,000 mark. Ten grand is a lot of money for a septic system. But if you don't have any choice, what can you do?

A chamber system is simple enough in design. Liquid leaves a septic tank and enters the first chamber. As more liquid is released from the septic tank, it is transferred into additional chambers that are farther downstream. This process continues, with the chambers releasing a predetermined amount of liquid into the soil over time. The process allows more time for bacterial action to attack raw sewage, and it controls the flow of liquid into the ground.

If a perforated-pipe system is used in ground where a chamber system is recommended, the result could be a flooded leach field. This might create health risks. It would most likely produce unpleasant odors, and it might even shorten the life of the septic field.

Chambers are installed between sections of pipe within the drain field. The chambers are then covered with soil. The finished system is not visible above ground. All the action takes place below grade. The only real downside to a chamber system is the cost.

Trench Systems

Trench systems are the least expensive version of special septic systems. They are comparable in many ways to a standard pipe-and-gravel bed system. The main difference between a trench system and a bed system is that the drain lines in a trench system are separated by a physical barrier. Bed systems consist of drain pipes situated in a rock bed. All the pipes are in one large bed. Trench fields depend on separation to work properly. To expand on this, let me give you some technical information.

A typical trench system is set into trenches that are 1 to 5 ft deep. The width of the trench runs from 1 to 3 ft. Perforated pipe is placed in these trenches on a 6-in bed of crushed stone. A second layer of stone is placed on top of the drain pipe. This rock is covered with a barrier of some type, to protect it from the backfilling process. The type of barrier used is specified in the septic system design.

When a trench system is used, both the sides of the trench and the bottom of the excavation are outlets for liquid. Only one pipe is placed in each trench. These two factors are what separates a trench system from a standard bed system. In bed systems, all the drain pipes are in one large excavation, and the bottom of the bed is the only significant infiltrative surface. Since trench systems use both the bottoms and the sides as infiltrative surfaces, more absorption is potentially possible.

Neither bed nor trench systems should be used in soils where the percolation rate is either very fast or slow. For example, if the soil will accept 1 inch per minute (in/min) of liquid, that is too fast for a standard absorption system. This can be overcome by lining the infiltrative surface with a thick layer (about 2 ft or more) of sandy loam soil. Conversely, drainage at a rate of 1 inch per hour (in/h) is too slow for a bed or trench system. In this situation, a chamber system might be recommended as an alternative.

Because of their design, trench systems require more land area than bed systems do. This can be a problem on small building lots. It can also add to the expense of clearing land for a septic field. However, trench systems are normally considered better than bed systems. There are many reasons for this.

Trench systems are said to offer up to 5 times more side area for infiltration to take place. This is based on a trench system with a bottom area identical to a bed system. The difference is in the depth and separation of the trenches. Experts like trench systems because digging equipment can straddle the trench locations during excavation. This reduces damage to the bottom soil and improves performance. In a bed system, equipment must operate within the bed, compacting soil and reducing efficiency.

If you have hilly land to work with, a trench system is ideal. The trenches can be dug to follow the contour of the land. This gives you maximum utilization of the sloping ground. Infiltrative surfaces are maintained while excessive excavation is eliminated.

The advantages of a trench system are numerous. For example, trenches can be run between trees. This reduces clearing costs and allows trees to remain for shade and aesthetic purposes. However, roots may still be a consideration. Most people agree that a trench system performs better than a bed system. When you combine performance with the many other advantages of a trench system, you may want to consider trenching your next septic system. It costs more to dig individual trenches than it does to create a group bed, but the benefits may outweigh the costs.

Mound Systems

Mound systems, as you might suspect, are septic systems which are constructed in mounds that rise above the natural topography. This is done to compensate for high water tables and soils with slow absorption rates. Because of the amount of fill material needed to create a mound, the cost is naturally higher than that for a bed system.

Coarse gravel is normally used to build a septic mound. The stone is piled on top of the existing ground. However, topsoil is removed before the stone is installed. A mound contains suitable fill material, an absorption area, a distribution network, a cap, and topsoil. Due to the raised height, a mound system depends on either pumping or siphoning to work properly. Essentially, effluent is either pumped or siphoned into the distribution network.

As the effluent is passing through the coarse gravel and infiltrating the fill material, treatment of the wastewater occurs. This continues as the liquid passes through the unsaturated zone of the natural soil.

The purpose of the cap is to retard frost action, deflect precipitation, and retain moisture that will stimulate the growth of ground cover. Without adequate ground cover, erosion can be a problem. There are a multitude of choices for acceptable ground covers. Grass is the most common.

Mounds should be used only in areas that drain well. The topography can be level or slightly sloping. The amount of slope allowable depends on the percolation rate. For example, soil that percolates at a rate of 1 in every 60 min or less should not have a slope of more than 6 percent if a mound system is to be installed. If the soil absorbs water from a percolation test faster than 1 in/h, the slope could be increased to 12 percent. These numbers are only examples. The professional who designs the mound system will set the true criteria for slope values.

Ideally, there should be about 2 ft of unsaturated soil between the original soil surface and the seasonally saturated topsoil. There should be a 3- to 5-ft-deep impermeable barrier. Overall the percolation rate could go as high as 1 in in 2 h, but this, of course, is subject to local approval. Percolation tests for this type of system are best done at a depth of about 20 in. However, they can be performed at shallow depths of only 12 in. Again, consult and follow local requirements.

How Does a Septic System Work?

How does a septic system work? A standard septic system works on a very simple principle. Sewage from a home enters the septic tank through the sewer. Where the sewer connects to the septic tank, there is a baffle on the inside of the tank. This baffle is usually a sanitary tee. The sewer enters the center of the tee and drops down through the bottom of it. The top hub of the tee is left open.

The bottom of the tee is normally fitted with a short piece of pipe. The pipe drops out of the tee and extends into the tank liquids. This pipe should never extend lower than the outlet pipe at the other end of the septic tank. The inlet drop is usually no more than 12 in.

The outlet pipe for the tank also has a baffle, normally

an elbow fitting. The drop from this baffle is frequently about 16 in.

When sewage enters a septic tank, the solids sink to the bottom and the liquids float within the confines of the container. As the tank collects waste, several processes begin to take place.

Solid waste that sinks to the bottom becomes what is known as *sludge*. Liquids, or *effluent* as it is called, are suspended between the lower layer of sludge and an upper layer of scum. The *scum* layer consists of solids and gases floating on the effluent. All three of these layers are needed for the waste disposal system to function properly.

Anaerobic bacteria work inside the septic tank to break down the solids. This type of bacteria is capable of working in confinements void of oxygen. As the solids break up, they form the sludge layer.

As the effluent level rises in the tank, it eventually flows out of the tank, through the outlet pipe. The effluent drains down a solid pipe to the distribution box. There the liquid is routed to different slotted pipes that run through the leach field.

As the effluent mixes with air, aerobic bacteria begin to work on the waste. These bacteria attack the effluent and eventually render it harmless. Aerobic bacteria need oxygen to do their job. Drain fields should be constructed of porous soil or crushed stone to ensure the proper breakdown of effluent.

As the effluent works its way through the drain field, it becomes odorless and harmless. By the time the effluent passes through the earth and becomes groundwater, it should be safe to drink.

Septic Tank Maintenance

Septic tank maintenance is not a time-consuming process. Most septic systems require no attention for years at a time. However, when the scum and sludge layers combined have a depth of 18 in, the tank should be cleaned out.

Trucks equipped with suction hoses are normally used to clean septic tanks. The contents removed from septic tanks can be infested with germs. The risk of disease due to exposure to sludge requires that the sludge be handled carefully and properly.

Drain field doesn't work

What happens if the drain field doesn't work? When a septic field fails to do its job, a health hazard exists. This situation demands immediate attention. The main reason for a field to fail in its operation is clogging.

If the pipes in a drain field become clogged, they must be excavated and cleaned or replaced. If the field itself clogs, the leach bed must be cleaned or removed and replaced. Neither of these propositions is cheap.

How can clogs be avoided?

Clogs can be avoided by careful attention to the types of waste entering the septic system. Grease, for example, can clog a septic system. Bacteria do not do a good job of breaking down grease. Therefore, the grease can enter the slotted drains and the leach field with enough bulk to clog them.

Paper, other than toilet paper, can also clog a septic system. If the paper is not broken down before entering the drain field, it can plug up the works.

Are garbage disposals harmful to a septic system?

Do garbage disposals hurt a septic system? The answers offered to this question vary from yes to maybe to no. Many people, including numerous code enforcement officials believe garbage disposals should not be used in conjunction with septic systems. Other people disagree and believe that disposals have no adverse effect on a septic system.

It is possible for the waste of a disposal to get into the

distribution pipes and drain field. If this happens, the risk of clogging is increased. Another argument against disposals is the increased load of solids they put on a septic tank. Obviously, the amount of solid waste will depend on the frequency with which the disposal is used.

I believe that disposals increase the risk of septic system failure and should not be used. However, I know of many houses using disposals with septic systems that are not experiencing any problems. If you check with your local plumbing inspector, this question may become a moot point. Many local plumbing codes prohibit the use of disposals with septic systems.

Piping Considerations

There are some additional piping considerations for plumbers to observe. Septic tanks are designed to handle routine sewage. They are not meant to modify chemical discharges and high volumes of water. If, as a plumber, you pipe the discharge from a sump pump into the sanitary plumbing system, which you are not supposed to do, the increased volume of water in the tank could disrupt its normal operation.

Chemical drain openers used in high quantities can also destroy the natural order of a septic tank. Chemicals from photography labs are another risk that plumbers should be aware of when piping drainage to a septic system.

Gas Concentrations

Gas concentrations in a septic tank can cause problems for plumbers. The gases collected in a septic tank have the potential to explode. If you remove the top of a septic tank when there is a flame close by, you might be blown up. Also, breathing the gases for an extended period can cause health problems.

Sewage Pumps

At times sewage pumps must be used to get sewage to a septic system. The pumps are normally installed in a buried box outside the building being served. The box is often made of concrete. In these cases, the home's sewer pipe goes to the pumping station. From the pumping station, a solid pipe transports the waste to the septic tank.

Sewage pumps must be equipped with alarm systems. The alarms warn the property owner if the pump is not operating and thus the pump station is filling with sewage. Without the alarm, the sewage could build up to a point where it would flow back into the building.

Sewage pumps have floats that are lifted as the level of the contents in the pump station builds. When the float is raised to a certain point, the pump cuts on, emptying the contents of the pump station. The discharge pipe from the pump must be equipped with a check valve. Otherwise, gravity would force waste down the pipe, back into the pump station, when the pump cut off. This would result in the pump having to constantly cut on and off, wearing out the pump.

An overflowing toilet

Some homeowners associate an overflowing toilet with a problem in the septic system. It is possible that the septic system is responsible for the toilet backup, but this is not always the case. A stoppage either in the toilet trap or in the drain pipe can cause a backup.

If you get a call from a customer with a flooding toilet, there is a quick, simple test that the homeowner can perform to tell you more about the problem. You know the toilet drain is stopped up, but will the kitchen sink drain properly? Will other toilets in the house drain? If other fixtures drain just fine, the problem is not with the septic tank.

There are some special instructions that you must give to customers prior to asking them to test other fixtures. First, it is best if they use fixtures that are not in the same bathroom as the plugged-up toilet. Lavatories and bathing units often share the same main drain that a toilet uses. Testing a lavatory that is near a stopped-up toilet can tell you if the toilet is the only fixture affected. It can, in fact, narrow down the likelihood of the problem to the toilet's trap. But if the stoppage is located some distance down the drain pipe, the entire bathroom group may be affected. It is also likely that if the septic tank is the problem, water will back up in a bathtub.

When an entire plumbing system is unable to drain, water will rise to the lowest fixture, which is usually a bathtub or shower. So if there is no backup in a bathing unit, there probably isn't a problem with a septic tank. But backups in bathing units can occur even when the major part of a plumbing system is working fine. A stoppage in a main drain could cause the liquids to back up into a bathing unit.

To find out whether a total backup is being caused, ask homeowners to fill the kitchen sink and then release all the water at once. Ask them to do this several times. A volume of water may be needed to expose a problem. Simply running the faucet for a short while might not reveal a problem with the kitchen drain. If the kitchen sink drains successfully after several attempts, it's highly unlikely that there is a problem with the septic tank. Thus the homeowner should call the plumber, not the septic installer.

Whole-house backups

Whole-house backups (where none of the plumbing fixtures drain) indicate a problem in the building drain, the sewer, or the septic system. There is no way to know where the problem is until some investigative work is done. It's possible that the problem is associated with the septic tank, but you will have to pinpoint the location of the trouble.

For all the plumbing in a house to back up, there must be some obstruction at a point in the drainage or septic system beyond where the last plumbing drain enters the system. Plumbing codes require cleanout plugs along drain pipes. There should be a cleanout plug either just inside or just outside the foundation wall. This cleanout location and the access panel of a septic tank are the two places to begin a search for the problem.

If the access cover of the septic system is not buried too deeply, start there. But if extensive digging is required to expose the cover, start with the cleanout plug at the foundation, hopefully on the outside of the house. Remove the cleanout plug and snake the drain. This will normally clear the stoppage, but you may not know what caused the problem. Habitual stoppages point to a problem in the drainage piping or septic tank.

Removing the inspection cover from the inlet area of a septic tank can show you a lot. For example, you may see that the inlet pipe doesn't have a tee fitting on it and has been jammed into a tank baffle. This could obviously account for some stoppages. Cutting off the pipe and installing the diversion fitting will solve this problem.

Sometimes pipes sink in the ground after they are buried. Pipes may be damaged when a trench is backfilled. If a pipe is broken or depressed during backfilling, there can be drainage problems. When a pipe sinks in uncompacted earth, the grade of the pipe is altered and stoppages become more likely. You might be able to see some of these problems from the access hole over the inlet opening of a septic tank.

Once you remove the inspection cover of a septic tank, look at the inlet pipe. It should be coming into the tank at a slight downward pitch. If the pipe is pointing upward, this indicates improper grading and a probable cause of stoppages. If the inlet pipe either doesn't exist or is partially pulled out of the tank, there's a very good chance that you have found the cause of the backup. If a pipe is hit with a heavy load of dirt during backfilling, it can be broken off or pulled out of position. This won't happen if the pipe is supported properly before backfilling, but someone may have cheated a little during the installation.

In a new septic system, a total backup is most likely to be the result of some failure in the piping system between the house and the septic tank. If the problem arises during very cold weather, possibly the drain pipe has retained water in a low spot and the water has since frozen. I've seen it happen several times in Maine with older homes.

Running a snake from the house to the septic tank will tell you if the problem is in the piping, assuming that the snake is a pretty big one. Little snakes might slip past a blockage that is capable of causing a backup. An electric drain cleaner with a full-size head is the best tool to use.

Problem in the tank

There are times, even with new systems, when the problem causing a whole-house backup is in the septic tank. These occasions are rare, but they do exist. When this is the case, the top of the septic tank must be uncovered. Some tanks, such as the one at my house, are only a few inches beneath the surface. Other tanks can be buried several feet below the finished grade.

Once a septic tank is in full operation, it works on a balance. The inlet opening of a septic tank is slightly higher than the outlet opening. When water enters a working septic tank, an equal amount of effluent leaves the tank. This maintains the needed balance. But if the outlet opening is blocked by an obstruction, water can't get out. This causes a backup.

Strange things sometimes happen on construction sites, so don't rule out any possibilities. It may not seem logical that a relatively new septic tank could be full or clogged, but it may be. Suppose a septic installer was using up old scraps of pipe for drops and short pieces, and one of the pieces had a plastic test cap glued into the end of it that was not noticed? This could certainly render the septic system inoperative once the liquid rose to a point where it attempted to enter the outlet drain. Could this really happen? I've seen the same type of situation happen with interior plumbing, so it could happen with piping at a septic tank.

What else could block the outlet of a new septic tank? Maybe a piece of scrap wood found its way into the septic tank during construction and is now blocking the outlet. If the wood floated in the tank and became aligned with the outlet drop, pressure could hold it in place and create a blockage. The point is, almost anything could be happening in the outlet opening; so take a snake and see if it is clear.

If the outlet opening is free of obstructions and all drainage to the septic tank has been ruled out as a potential problem, look farther down the line. Expose the distribution box and check it. Run a snake from the tank to the box. If it comes through without a hitch, the problem is somewhere in the leach field. In many cases, a leach field problem will cause the distribution box to flood. So if liquid is rushing out of the distribution box, be alert to a probable field problem.

Problems with a Leach Field

Problems with a leach field are uncommon in new installations. Unless the field was poorly designed or installed improperly, there is very little reason for it to fail. However, extremely wet ground conditions, due to heavy or constant rains, could make a field saturated. If the field saturates with groundwater, it cannot accept the effluent from a septic tank. This, in turn, causes backups in houses. When this is the case, the person who created the septic system design may be to blame.

Older fields

Older fields sometimes clog up and fail. Some drain fields become clogged with solids. Financially, this is a devastating discovery. A clogged field has to be dug up and replaced.

Much of the crushed stone might be salvageable, but the pipe, the excavation, and whatever new stone is needed can cost thousands of dollars. The cause of a problem of this nature is a poor design, bad workmanship, or abuse.

If the septic tank installed for a system is too small, solids are likely to enter the drain field. An undersized tank could be the result of a poor septic system design, or it could come about as a family grows and adds to the home. A tank that is adequate for two people may not be able to keep up with the usage seen when four people are involved. Unfortunately, one doesn't find out that a tank is too small until the damage has already been done.

Why would a small septic tank create problems with a drain field? Septic tanks accept solids and liquids. Ideally, only liquids leave the septic tank and enter the leach field. Bacterial action occurs in a septic tank to break down solids. If a tank is too small, there is not adequate time for the breakdown of solids to occur. Increased loads on a small tank can force solids down into the drain field. After this happens for awhile, the solids plug up the drainage areas in the field. Then digging and replacement are needed.

Is there such a thing as having too much pitch on a drain pipe? Yes, a pipe that is graded with too much pitch can cause several problems. In interior plumbing, a pipe with a fast pitch may allow water to race by without removing all the solids. In a properly graded pipe, the solids float in the liquid as drainage occurs. If the water is allowed to rush out, leaving the solids behind, a stoppage will eventually occur.

In a septic tank, a pipe with a fast grade can cause solids to be stirred up and sent down the outlet pipe. When a 4-in wall of water dumps into a septic tank at a rapid rate, it can create quite a ripple effect. The force of the water might generate enough stir to float solids that should be sinking. If these solids find their way into a leach field, clogging is likely.

We talked a little bit about garbage disposals earlier. When a disposal is used in conjunction with a septic system, more solids are involved than there would be without a disposal. Where code allows, this calls for a larger septic tank. Because of the increase in solids, a larger tank is needed for satisfactory operation and reduced risk of a clogged field. Remember, some plumbing codes prohibit the use of garbage disposals where a septic system is present.

Other causes of field failures can be related to collapsed piping. This is not common with today's modern materials, but it is a fact of life with some old drain fields. Heavy vehicular traffic over a field can compress it and cause the field to fail. This is true even of modern fields. Saturation of a drain field will cause it to fail. This could be the result of seasonal water tables or prolonged use of a field that is dying.

The solids in septic tanks should be pumped out on a regular basis. For a normal residential system, pumping once every 2 years is adequate. Septic system professionals can measure sludge levels and determine whether pumping is needed. Failure to pump a system routinely can result in a buildup of solids, which may invade and clog a leach field.

Normally, septic systems are not considered a plumber's problem. Once you establish that a customer's grief is due to a failed septic system, you should be off the hook. Advise the customer to call septic system professionals, and go to your next service call; you've earned your money.

10

Remodeling Reminders

This chapter is chock full of remodeling reminders. Plumbers who are not accustomed to doing remodeling work are often caught between a rock and a hard place. They don't have the experience to know what to look for when they get into remodeling. And the risks can be substantial. Remodeling work can be very different from new work or service work. For an unsuspecting plumber, a few remodeling jobs is all it takes to lose enough money to put a real strain on the business.

There is no doubt that remodeling can be hard work. It can also be dangerous work. And the work can be much more extensive than it first appeared. This fact often leads plumbers to give bids and prices that are far lower than they should be. Remodeling is a good field to get into. It survives when new work dwindles, and remodeling can be done throughout the year. Snow and cold weather don't shut down remodelers. If you are thinking of getting into remodeling, read this chapter first!

Pipe Sizes

One of the first considerations when you are adding to an existing system is the size of existing pipes. Before you can create a plumbing design, you must inspect the existing pipes that you will be tying into. Check the size of the sewer where it enters the home. If it is a 3-in pipe, you could be in for big trouble. A 3-in sewer must not carry the discharge of more than two water closets. If you plan to install a third water closet on a 3-in sewer, you are out of luck. Before the job can be done, you will have to install a new sewer. It may be more cost-effective to run a new sewer, independent of the old sewer, for the new plumbing. In most cases, it makes more sense to replace the existing sewer with a larger one. Before you become too involved in your remodeling job, make sure the sewer is large enough to carry the increased fixture-unit load.

As with the sewer, check the size of the incoming water service. Verify the ability of the water service to provide an adequate water supply. Rarely will you have to replace the water service with a larger one.

Water distribution pipes must be inspected to ensure adequate sizing. Notice the type of pipe used in the water distribution system. Since you have to connect new plumbing to these existing pipes, it helps to know the type of pipe used.

A common problem with older homes is the size of the water pipes, especially the hot water pipes. It is not unusual to find homes where all the hot water distribution is done through ½-in pipes. As you may know, ½-in pipe is not large enough to supply the hot water for a house. If you cannot find ¾-in water pipes to connect to, you may have to apply for a variance. Generally, if the only water pipes in the house are ½-in, the code enforcement officer will allow you to connect to them for the new installation. While this is a common practice, don't count on it. The code officer can require you to run the proper-size piping for the installation. If this happens, you may have to bring the remainder of the home's plumbing up to current code requirements. If at any time you have questions about your responsibilities concerning the plumbing code, consult your local plumbing inspector.

The Water Heater

With average remodeling jobs, the existing water heater will be sufficient. However, if you are installing a whirlpool tub or anything that will demand large amounts of hot water, consider the possibility of adding a new water heater. It is your responsibility to install plumbing that works well and that conforms to local plumbing code.

Old Plumbing That Must Be Moved

It is easy to get caught up in figuring out how to install new plumbing, without considering what may need to be moved in the existing system. If an attic is being converted to living space, the existing vents may have to be relocated. Should existing walls be removed, there may be concealed plumbing to move. It could be water pipes, drains, or vents. This is a potential task that can be difficult to foresee.

When you discuss the overall remodeling plans with a general contractor or owner, consider wall removal. If a wall is to be removed, any pipes in the wall will have to be relocated. When a wall is scheduled for demolition, check the attic and the basement or crawl space for pipes entering or exiting the wall. If you see pipes that will have to be moved, note the type of pipe. Determine the type of material used in the pipe and the purpose of the pipe. This information will help you to design a plan for the successful relocation of the pipe.

If the existing water distribution system is piped with galvanized-steel pipe, give serious consideration to replacing all the galvanized pipe. Galvanized-steel pipe is famous for rusting from the inside out. As the pipe rusts, two conditions develop that affect the water distribution system.

The first condition is restricted water flow in the pipe. As the pipe rusts, the rough surface collects minerals and other undesirable objects which restrict the water flow. In time, the pipe can become so closed that only a dribble of water comes out of the faucets. The second condition of rusting galvanized-steel water pipes is the development of leaks. Where the pipe has been threaded, the wall of the pipe is thinner. The threading process weakens the pipe, and rust attacks these weak spots. As the rust spreads, the pipe threads deteriorate and leak. Unless there are extenuating circumstances, replace all the galvanized-steel water pipe you can.

On numerous occasions I have found existing plumbing systems piped with illegal materials. The most common violation in the water distribution system is the use of polyethylene pipe for the entire water distribution system. Since this pipe is not approved for use with hot water, it cannot be used as a water distribution pipe in houses with hot water. The cold water pipes must be made from the same material used for the hot water pipes.

Fixtures

When it is your job to replace existing fixtures, prepare for the unexpected. Working with old fixtures can turn your hair gray, quickly. Old plumbing fixtures often entail some undesirable working conditions. The list begins with rusted mounting nuts on faucets and includes everything from broken, or nonexistent, closet flanges to odd-size bathtubs.

If you have never tried to remove a double-bowl, concrete laundry tub, you cannot imagine how heavy it is. Cast-iron bathtubs are another heavy fixture. Unless you have a small army to help you, plan on breaking up these accidents-waiting-to-happen with a sledge hammer. The danger inherent in moving such heavy fixtures outweighs the loss incurred by destroying the fixtures with a hammer. Having a 400-lb cast-iron tub chase you down a set of stairs will convince you that I am right.

Galvanized-steel drains

Galvanized-steel drains have earned quite a reputation in the world of pipe blockages. Like galvanized-steel water pipes, galvanized-steel drains are proven to rust. In drains, the inside roughness, caused by rust, catches hair, grease, and other unidentified cruddy objects. As these items become snagged on the rough spots, the interior of the drain slowly closes. In time, the buildup will block the passageway of the pipe entirely. The result is a drainpipe that will not drain.

When you try to fix this type of blockage with an average snake, the snake only punches a hole in the blockage. The drain works for awhile, but before long the pipe is stopped up again. Unless the blockage is removed with a cutting head on an electric drain cleaner, the pipe will become restricted soon after the snaking.

When a garbage disposal is added to a kitchen sink with a galvanized-steel drain, expect problems. Since garbage disposals send food particles down the drain, rough spots in the drain will catch and hold the food. You can almost count on having to replace any old galvanized-steel pipe used in a drain for garbage disposals. In addition to the rust problem, the fittings used with galvanized-steel pipe are not as effective as modern fittings. New fittings utilize longer turns. A galvanized-steel quarter bend takes a much sharper turn than a schedule 40 plastic quarter bend. These tight turns are responsible for creating stoppages in the pipe.

In addition to habitual stoppages, galvanized-steel drains will develop leaks at their threads. This is caused by rust working on the weak spots at the threads. In general, galvanized-steel pipe should be replaced whenever feasible.

Septic Systems

If you are adding plumbing to a house that uses a septic system, be careful. The septic system may not be adequate to handle the increased demand from new plumbing. Before you add plumbing to an old septic system, have the system checked out. You may be able to determine the system's capability by reviewing the original installation permit and plans. If this paperwork is not available, hire a professional to render an opinion on the ability of the system to handle the increased load.

Septic systems are expensive to install and expand. It would be a shame to add a new bathroom, only to find that the septic system was not adequate to process the waste. The cost of adding new chambers or lines in the septic field may shock you. Although you may not be happy to discover that you need to invest extra money in the septic system, it is better to find this out before the new plumbing is installed.

Rotted Walls and Floors

When you plan to replace old plumbing fixtures, you can never be sure what you will find during the process. It is not uncommon for toilets to leak and cause the floor beneath them to rot. Bathing units are subject to the risks of rotten floors and walls. These structural problems will complicate your life as a plumber.

When an old toilet is removed and a rotten floor is found, someone will have to pay to have the floor repaired. If you are the homeowner, that someone is you. Toilets contribute to the rotting of floors through leaks around wax seals, tank-to-bowl bolts, and condensation. While the quantity of water running onto the floor may be minor, the damage can be major. As time passes, the water works its way into the underlayment, subfloor, and floor joists. Small leaks may not stain the ceiling below a toilet until the damage already has been done.

If you are concerned about water damage around the base of the toilet, test for it before you remove the old toilet. Use a knife or screwdriver to probe the floor around the base of the toilet. If the point of the probe sinks into the floor, you have a problem. If the water damage is only in the underlayment or subfloor, the cost of repair will be minimal. If you were not already planning to replace the floor covering, you will incur additional expenses. To correct the problem, the finished floor covering has to be replaced.

Trying to make a patch in the floor covering will just result in a floor that does not match.

Bathtubs and showers may have caused hidden damage to the floor or walls. When people step out of the bathing unit, they often drip water on the floor. If the seal at the base of the tub or shower is not good, this water runs under the bathing unit. Prolonged use under these conditions results in severe water damage to the floor and floor joists.

If the walls surrounding the bathing unit leak, water will invade the wall cavity. Water may rot the studs or the base-plate of the wall. Given enough time, water will penetrate the wall's plate and enter the floor structure. This type of damage may go unnoticed until you remove the old fixtures. When the damage is found, it will have to be repaired before you can proceed.

The structural damage caused by water can be substantial. I have seen floors so rotted that the fixtures fell through the floor. It is not only embarrassing to have your toilet or tub fall through the floor; it is dangerous. If you will be responsible for the costs incurred to repair water damage, allow for the possibility in your remodeling budget.

Cutting into Cast-Iron Stacks

When you are remodeling, it is not unusual to work with cast-iron drains and vents. Cutting these pipes to install a new fitting can be very dangerous. Many vertical pipes are not well secured. When you cut the pipe to install the new fitting, the pipe above you may come crashing down. This is a potentially fatal situation.

If you must cut into a vertical, cast-iron stack, be sure that it will not fall when cut. Find a hub on the pipe, and use perforated strap to secure the pipe. Wrapping the galvanized-steel strapping around the pipe, under the hub, and nailing it to the studs or floor joist will usually protect you. Whenever possible, have a helper on hand when you cut the stack. The method used to secure the pipe will vary with individual circumstances, but be sure the pipe will not move when it is cut.

Adding a Basement Bathroom

Putting a bathroom in the basement can be a problem when it comes time to vent it. The bathroom must have a vent that goes up into the living space above the basement. Either take the vent through the roof, or tie it into another suitable vent. If you are tying into an existing vent, tie in at least 6 in above the flood-level rim of the fixture served by the existing vent.

Size Limitations on Bathing Units

When adding a bathroom in an existing house, you must be aware of access limitations. The size of the ingress areas dictates the maximum size of the fixtures. For example, one-piece, fiberglass tub/shower combinations will not fit through the doors and stairways of most homes. One-piece units are used almost exclusively in new construction and additions. Under new-construction conditions, these bathing units can be set in place before walls and doors are installed.

Even if you are able to get one of these units in a front door, it probably will not go around turns or up stairways. To solve this problem, look at sectional bathing units. Many types of sectional units are available. There are two-piece, three-piece, four-piece, and five-piece units on the market. With so many to choose from, you should have no difficulty finding one to suit your needs.

Cutoffs That Don't

Remodeling jobs are frequently complicated by valves that will not close. In these cases, you may have to cut off the water at the water meter. Never assume that an existing valve will function properly. Cut off the valve and confirm that the water is off by opening a faucet. Many old valves become stuck and will not close. Even when the handle turns, the valve may not be closing. If you fail to double-check the effectiveness of the valve, you may flood the home when you cut pipes or loosen connections.

Working with Old Pipes

Working with old pipes can complicate the life of a plumber. However, modern devices have made the job of coupling two different types of pipe much easier. Let's take drain-waste-and-vent (DWV) copper pipe as an example. There are still a large number of plumbing systems in operation that were constructed from DWV copper. This thin-walled copper has a long life as a drain and vent material. When you find DWV copper, it is usually in good working condi- tion. Since this pipe creates few problems, there is no need to replace it. However, it is rarely cost-effective to use cop- per to extend a system in today's plumbing practices. When you want to add to a system configured from copper, you will probably connect to the copper with plastic pipe.

When your goal is to mate plastic pipe with copper pipe, you have a few options. The most common method for con- verting copper to plastic uses threaded connections or rub- ber couplings. Either of these adapters will give satisfacto- ry results, without much effort or expense.

The first step in connecting to an existing copper pipe is to cut it. Make room for your tee or wye. Roller-type cutters are the best choice for cutting copper, if you have room to operate them. When space is limited, you can cut the cop- per pipe with a hacksaw or a reciprocating saw, fitted with metal cutting blades. The principles for soldering DWV copper are the same as those used for copper water pipe.

When you cut in a copper fitting, you must consider how much pipe to remove for the fitting. Another factor is the flexibility of the existing pipe. If the existing pipe cannot be moved forward, backward, or vertically, use an additional fitting to make the connection. Once the pipe is cut, pre- pare and install the new fitting. It is important that all the pipes entering the new fitting seat properly. This is not to say that the pipe must be jammed into the fitting to the hilt; but there must be enough pipe in the fitting to ensure a solid joint. If the existing pipe has very little play, you may not be able to install the new fitting without a slip coupling.

Slip couplings do not have the ridges near their center-points, as standard couplings do. A slip coupling can slide down the length of a pipe, without being stopped by the ridges found in normal couplings. This sliding ability makes a slip coupling indispensable for some adaptation work. To install a normal fitting, you must have room to move existing pipes back and forth. With slip couplings, this space is not required.

If you have to use a slip coupling, cut additional pipe out of the existing system to install the new fitting. The new fitting goes on one end of the existing pipe, and the slip coupling is slid back on the other end of the existing pipe. Put a new piece of pipe in the vacant end of the new fitting. The pipe has to be long enough to reach nearly to the end of the existing pipe. Then slide the slip coupling back toward the new pipe. Thus you can connect the two pipes without the need to move existing pipes.

Once the branch fitting is installed, you are ready to install the threaded adapter, which may have male or female threads. Install a piece of pipe between the branch fitting and the adapter. These joints are soldered in normal practice. When the pipe and fittings have cooled, you are ready to switch pipe types.

When the copper has cooled, apply pipe dope to the threads of the male adapter. The male adapter may be copper or plastic, depending upon which order you have chosen for the adapters. Screw the plastic adapter into or onto the copper adapter. When the plastic adapter is tight, the conversion has been made. From this point, you can run plastic pipe in the normal manner.

If you wish there were an easier way to convert DWV copper to schedule 40 plastic pipe, there is. Rubber couplings can simplify the conversion between the two pipe types. When you use rubber couplings, there is no need to solder. The amount of pipe you cut out of the existing system is not critical when you use rubber couplings. The measurements are not critical, because you can bridge the span with plastic pipe. Rubber couplings can slide up the existing pipes, like a slip coupling, so pipes that cannot move are no problem. All aspects considered, I can see no reason to use any other type of conversion method in normal applications.

The first step is to cut a section of pipe out of the existing plumbing system. Remove enough pipe to allow the installation of the branch fitting and two pieces of pipe. With the old pipe removed, slide a rubber coupling on each end of the existing pipes. Hold your branch fitting in place, and measure the two pieces of pipe that will join the fitting to the existing pipes. Install the two pipes in the ends of the branch fitting.

Hold the branch fitting, and its pipe extensions, in place to check your measurements. The plastic pipes protruding from the branch fitting should extend to a close proximity to the copper pipes. With the pipes in place, slide the rubber couplings over the plastic pipes. Position the stainless-steel clamps in the proper locations, and tighten the clamps. When the clamps are tight, the conversion is complete. Now you can work directly from the plastic branch fitting to complete the installation.

Galvanized-steel drains

Many houses are still equipped with galvanized-steel drains and vents. Because of galvanized-steel pipe's potential problems, it is a good idea to replace galvanized-steel pipes whenever possible. When you do not replace the pipe, use adapters to make the steel pipe compatible with new plumbing materials.

In many situations you have to cut the old galvanized-steel piping to install a branch fitting for the new plumbing. A hacksaw is the tool most often used to accomplish this task. A reciprocating saw with a metal-cutting blade makes the job easier. When you use an electric saw, be careful not to electrocute yourself. The saw must be insulated and designed for this type of work. Old drains often hold pockets of water; if you hit this water with a faulty saw, you will get a shock.

Rubber couplings are the easiest way to convert galvanized-steel pipe to a different type of pipe. If you elect to use rubber couplings, the procedure is about the same as that described for copper pipe. Cut out a section of the galvanized-steel pipe to accommodate the new branch fitting. Then install the branch fitting and its pipe sections with rubber couplings. The couplings simply slide over the ends of each pipe and are held in place by the stainless-steel clamps.

Since you cannot solder a fitting onto galvanized-steel pipe, if you choose to use threaded adapters, you must have threads available to work with. When cutting in a branch fitting, you must cut the old pipe and remove it until you get to a threaded fitting or pipe end. This can be difficult. The old pipe may have concealed joints, or the pipe may have seized in its fittings. Pipe wrenches are a must for disassembling old galvanized-steel drains.

Once you have worked your way to a threaded fitting or an end of the pipe with threads on it, install the threaded adapters. Apply pipe dope to the male threads of the connection, and screw the adapter into or onto the threads. Tighten the adapter with a pipe wrench. When the adapters are tight, proceed with the new piping material.

When you cut in a branch fitting on galvanized-steel pipe, rubber couplings are the easiest. The rubber couplings work fine and reduce the time and frustration involved in working with old galvanized-steel pipes. If you are only converting a trap arm, or similar type of section, threaded adapters are a reasonable choice. With a trap arm, you unscrew only one section of the steel pipe. Once it is removed, you have the threads of the fitting to screw an adapter into. Under these conditions, threaded adapters are the best choice.

Cast-Iron Pipe

Cast-iron pipe has been used for many years to provide drains and vents in plumbing systems. Even today, cast iron is still used for these purposes. For the most part, cast iron has been replaced with plastic pipe in modern systems, but cast iron does still see some use in modern systems. There are two types of cast iron that you are likely to encounter. The first type is a service-weight pipe with hubs. This pipe is often called *service-weight cast iron* or *bell-and-spigot cast iron*. The second type is a more lightweight pipe that does not have a hub.

Service-weight cast iron is what you are most likely to encounter. This pipe has been used for a long time and is still used. It is the pipe used in most older homes. Service-weight cast iron is usually joined with caulked lead joints at the hub connections. Although caulked lead joints were used to install this pipe, modern adapters can be used to avoid working with molten lead. The options for using adapters with bell-and-spigot cast iron are discussed in the following paragraphs.

The lightweight, hubless cast-iron pipe used in some jobs is a much newer style of cast iron. The connections used for this type of pipe are a type of rubber coupling. They are not the same rubber couplings referred to throughout this chapter. These special couplings are designed for joining hubless cast iron. The couplings have a rubber band that slides over the pipe and fitting. There is a stainless-steel band that slides over the rubber coupling. The band is held in place by two clamps. When the clamps on the band are tight, they compress the rubber coupling and make the joint. These special couplings are easy to work with, but they are not the only option. The heavy rubber couplings discussed elsewhere in this chapter will work fine with this pipe.

Cast-iron pipe can be cut with roller cutters and metal-cutting saw blades, and some people can even cut it with a hammer and chisel. For the average job, rachet-style snap cutters are best for cutting cast-iron pipe. These tools can be used to cut pipe when there is very little room to work with. If you can get the cutting chain around the pipe and have a little room to work the handle back and forth, you can cut the pipe quickly and easily. These soil-pipe cutters are expensive, but you can rent them.

Rubber couplings

The heavy rubber couplings discussed earlier will work fine with either type of cast-iron pipe. They are used in the same manner as described earlier. When working with hubless cast iron, you can use the special rubber couplings designed for hubless pipe. If you use these special couplings to mate cast iron to plastic, you will need an adapter for the plastic pipe. It is possible to use the couplings without the plastic adapter, but you shouldn't. The plastic adapter has one end formed to the proper size to accommodate the special coupling. The adapter is glued onto the plastic, and then the special coupling slides over the factory-formed end of the adapter.

Cast-iron doughnuts

When you are working with fittings for service-weight cast iron, special adapters allow plastic pipe to be installed in the hub. Normally, when pipe is placed in the hub of service-weight cast iron, it is held in place by a hot-lead joint. To avoid using molten lead, you can use a ring adapter. Most plumbers call these ring adapters *doughnuts*.

These ring adapters are not always easy to use. They fit in the hub of the cast iron, and the new pipe is driven into the ring. As the new pipe goes into the ring, the ring grips the pipe and forms a watertight joint. The problem arises when you are trying to get the pipe into the ring. Lubricate the pipe and the inside of the ring generously before you attempt the installation. Special lubricant is made for the job and is sold by plumbing suppliers.

Place the doughnut in the hub of the fitting. Apply plenty of lubrication to the inside of the ring and to the outside of the pipe. Place the end of the pipe in the ring and push.

The pipe will probably not go very far. When the pipe is as far in the ring as it will go, use some force. Take a block of wood and place it over the exposed end of the pipe. Hit the wood with a hammer to apply even pressure to the pipe. With some luck and a little effort, the pipe will be driven into the doughnut. Once the pipe is all the way into the adapter ring, proceed with the new piping in the usual manner.

Copper Water Pipes

A vast majority of today's homes have copper pipe installed for the distribution of potable water. If you do not wish to make the new installation in the same type of pipe, you have to use adapters. When you prefer to work with plastic pipe, convert the copper to a suitable pipe type. This can be done with a variety of adapters. Of all the adapters available, threaded adapters are the most universal.

To cut copper pipe is not difficult. The job is done easiest with roller cutters, but it can be accomplished with a hacksaw. In remodeling, miniature roller cutters often come in handy. Measuring to determine how much pipe to remove for a branch fitting is done as described for DWV copper.

Threaded adapters

By using threaded adapters, you can mate copper water pipes with any other approved material. Cut in the tee and solder the threaded adapter to a piece of copper at the point where you wish to make the conversion. After the soldered joint cools, apply pipe dope to the male threads to be used in the connection. Screw the male threads into the female threads until the fitting is tight. Now the conversion is complete. Continue the new installation from the opposite end of the threaded adapter.

Compression fittings

Compression fittings are another possibility for converting copper to a different type of pipe. Take a compression tee and cut it into the copper pipe. The same measurement techniques used for copper fittings work with compression fittings. Since the branch fitting is of the compression type, there is no soldering to be done.

Saddle valves

Self-piercing saddle valves are a good choice when you are adding new plumbing for an ice maker or similar appliance. The saddle clamps to the copper pipe and allows the use of plastic or copper tubing between the saddle and the appliance. The saddle valve connects to the tubing with a compression fitting. The saddle is held in place on the copper water pipe with a clamp-and-bolt device. With a self-piercing saddle valve, you do not have to drill a hole in the copper water pipe. When the handle on the saddle is turned clockwise to its full extent, the copper water pipe is pierced. When the handle is turned back counterclockwise, the water flows into the tubing.

Painted Copper Pipes

The copper water pipes in older homes may be covered in paint. At one time, it was fashionable to paint exposed pipes to blend in with the decor of the home. The paint may have made the pipes more attractive, but it definitely makes soldering these pipes more difficult. If you are required to cut a branch fitting into painted pipes, remove the paint from the area to be soldered.

Most plumbers first cut the painted pipes and then try to remove the paint. They usually use regular sanding cloth and struggle to get the paint off. There is an easier way to remove old paint. Remove it *before* you cut the pipes. Before the pipes are cut, they are solid, and it is easy to apply pressure to them with the sandpaper. Once the pipes are cut, they wobble and bounce around, making it difficult to apply steady pressure with the sanding cloth. Another common problem in cutting pipes before the paint is removed is the water in the pipes. When the pipes are cut, water often runs down the pipe, wetting the paint. The wet pipe gets the sanding cloth wet. When the pipe and the sandpaper are wet, the paint is much harder to remove.

The standard sanding cloth used for copper pipe removes most paint, but you have to sand the pipe several times. If you know you will be working with painted pipes, put some steel wool in your tool box. Also invest in some sandpaper with a heavy grit. The rougher sandpaper will do a better job than the fine-grit sandpaper usually used with copper. If you use steel wool and coarse sandpaper before the pipes are cut, the paint should come off with minimal effort.

If the paint does not come off, burn it off. If you have enough clear space to avoid an unintentional fire, use a torch to burn the paint off the pipes. When flammable materials are nearby, you must protect against an accidental fire. You can reduce the risk of unwanted fire in many ways. Use a heat shield on the torch, if you have one. If you don't have a heat shield, you can make one from ordinary household goods. By folding aluminum foil into several layers, you can make a reasonable heat shield. A piece of sheet metal also serves as a heat shield.

Spraying water on flammable materials reduces the risk of accidental fires. For best results, wet down all flammable materials. Place a heat shield between the flame and any flammable materials. Have an adequate fire extinguisher on hand, just in case. Proceed cautiously with your work, keeping an alert eye on the flame and the flammable materials. If the flammable material starts to smoke, turn black, or bubble, cut off the torch and dowse the hot material with water. If an unwanted fire has progressed beyond the early stage, quickly put it out with a fire extinguisher. If you use a heat shield and wet down the surrounding materials, you should not have a problem with unwanted fire.

After you remove the paint, cut the pipes. But before you cut the pipes, be sure the water is off. Drain as much water from the pipes as possible. If you can see water standing in the pipe, you know soldering will not be easy. It is not unusual for water to remain in pipes after they are cut. If the pipes run horizontally, you may be able to pull down on the pipe to allow the water to escape. When working with a vertical pipe, you won't be able to tip it down to drain it. Instead, remove the water with a drinking straw. Don't worry, you won't have to suck the water out of the pipe. Place the straw in the vertical pipe. When the straw is in the pipe, blow through it. The pressure forces the standing water up and out of the pipe. Unless water is leaking past a valve, this will clear enough water for you to do the soldering.

Apply a generous amount of flux to the pipe where the fitting will be placed. Before you install the fitting, heat the flux with the torch. As the flux bubbles and burns, it cleans the pipe. After this cleaning, follow usual soldering procedures to complete the job.

Remodeling work can be very different from new work. If you don't have experience in remodeling, be careful when giving prices to customers. A job that looks easy can turn into a major project. Assess all conditions carefully, and proceed with caution. If you do this, you should gain the experience you need with a minimal loss of profit.

Clogged Drains

Clogged drains are a big bother for homeowners, but they can produce constant bread-and-butter money for plumbers. Contractors who are geared up to handle drain-cleaning jobs can earn a lot of money in a very short time. This is especially true when they use flat-rate pricing schedules. A typical drain that may take less than 30 minutes to clear can earn a profit for a plumber that is equal to 90 minutes of labor charges. Part of this is due to the flat-fee pricing structure. Another part is a machine rental charge that is billed to customers.

In cities, where the population warrants it, many plumbing contractors specialize in drain cleaning. There are franchises that are built around drain cleaning. In many jurisdictions, a plumber's license is not required to clean drains. This means that some unlicensed people can cut in on the territory that once belonged to plumbers. It also means that plumbing contractors can use trained, but unlicensed, people to take service calls for clogged drains. The profit potential from doing this is phenomenal.

Drain cleaning is not a part of every plumber's career, but it is a specialty for some plumbing contractors. There is big money to be made in drain cleaning, and this big (and often easy) money attracts a lot of plumbers. Dealing with overflowing toilets and backed-up drains may not be your idea of good working conditions, but some contractors wouldn't trade it for any other type of work.

Why is this type of work so appealing to those who like it? There are several reasons, but money is the major one. If I go out and replace a ball cock, I'm going to get paid for one hour's labor and the parts I use. There will be some markup on the parts, but it won't amount to much.

If I go out to clear a blockage from a building drain or sewer, I'm still going to get my hourly labor rate, but I'm also going to charge a rental fee for the use of my drain-cleaning machine. This fee could run anywhere from $25 to $65. When you couple an hourly rate with a rental charge, the total is attractive to a plumber's profit-and-loss statement.

Once a sewer machine is paid for, the rental fee is gravy. When the call comes in late at night, the labor and rental fees are going to be much higher. It is simple economics; plumbers make more money when they charge a rental fee on drain-cleaning equipment.

Another reason for some plumbers to specialize in drain cleaning is the lack of inventory needed. A typical service van can easily be stocked with an inventory worth more than $10,000. This is a lot of money to tie up for each truck on the road. If all a plumber is doing is drain cleaning, the inventory needs cost less than $500. If you had five service vans on the road, this would make quite a difference.

The contents of five full-service plumbing vans might cost $50,000, not counting the cost of tools and equipment—which would include drain-cleaning equipment. Five trucks could be put on the road to specialize in drain cleaning with only $2500 in inventory. That leaves $47,500 to use for advertising, reserve capital, etc. Are you starting to see the attraction of drain cleaning?

There are, of course, other reasons why some plumbers prefer drain cleaning to other types of plumbing. It is usually feasible for one person to handle drain-cleaning jobs. A one-plumber shop will have trouble doing large commercial jobs. Working alone can be tough in any type of new construction or remodeling plumbing. Service work and drain cleaning both offer opportunity to the individual owning a business without employees.

Drain cleaning is often a fast process when proper equipment is used. This makes it possible to get in and out of a job quickly. If you are working on a flat-rate first-hour charge, this is advantageous.

Many homeowners try to fix their own minor plumbing problems, but few will go to a rental center and rent a drain cleaner to auger out the drains. Thus there is lots of work for contractors specializing in drain cleaning. Even when homeowners attempt to snake their own drains, they usually fail.

When a toilet is flooding a bathroom or sewage is backed up in a bathtub, people don't put off calling a plumber. They might live with a dripping faucet or less hot water than they would like, but when the drains stop up, people call for help. This is another advantage for the drain-cleaning plumber.

As you can see, there are a number of reasons why drain cleaning is worthy of consideration as a part of your career. Certain dangers are inherent in drain cleaning. There are also right and wrong ways to do the job. This chapter talks about these facets of the job, and introduces the tools of the trade for drain cleaning. You will also be told how, why, and where various methods of drain cleaning are used.

Liquid and Powder Drain Openers

Liquid and powder drain openers that are sold in the grocery store rarely work on serious pipe stoppages. These drain openers can cut through a limited amount of grease and hair, but when the clog is tough, these over-the-counter products won't cut it.

While most of these homeowner-oriented liquids and powders won't solve the drainage problem, they do create a problem of their own for you, the plumber. These drain cleaners can be dangerous to your skin, eyes, and respiratory system. When you respond to a call for drain cleaning, ask whether any chemicals or drain cleaners have been added to the backed-up pipe or fixture. It is a good idea to wear gloves, eye protection, and even a mask when you clear drain stoppages.

As a professional plumber, you have access to some liquid drain openers that do work. Most of these liquids have a high concentration of acid. Don't handle these materials carelessly. Also be aware that combining an acid-based product with a lye-based product can result in an eruption of harmful chemicals.

Personally, I don't like liquid drain cleaners. There are many times when even the professional-strength varieties don't work. I prefer to clean drains the old-fashioned way—with mechanical devices.

Water-Powered Drain Openers

Water-powered drain openers can flush out a pipe quickly. However, when a drain is vented, getting these water-powered devices to work can be a problem. Many plumbers call these devices *sewer bags*. They are either canvas or rubber devices that connect to the threads of a garden hose. The device is placed in the pipe containing the stoppage, and the water to the hose is turned on. The bag swells and seals off the opening of the clogged pipe. Water pushes out of the bag, building volume and pressure, until the clog is blown out. Normally, the water pressure breaks up the clog or pushes it to a larger pipe where it can be drained from the system. However, on some occasions the pressure blows the sewer bag out of the pipe. These occasions are rare, but they do occur.

A sewer bag will work only if the water being pumped into the drain is confined between the clog and the sewer bag. If the water can rise up a vent or come out in a fixture that connects at some point to the drainage system between the bag and the clog, it will. I learned this the hard way.

Air-Powered Drain Openers

Air-powered drain openers follow the same principles as sewer bags. Air is forced into the drainage system to blow out the clog. These devices are also rendered useless if air is allowed to escape up a vent. Kinetic water rams are the professional version of air-powered drain openers, but many homeowner versions are available, including plungers.

Closet Augers

Closet augers are used exclusively for clearing stoppages in the traps of water closets. Their design allows them to be snaked through the integral trap of a toilet without damaging the china, in most cases.

Most professional augers have a spring cable that is about 12 ft long. The cable is retracted into the sleeve of the auger until only the big spring head extends past the curved section of the sleeve. The curved section is set in the bottom of the toilet bowl and is positioned toward the trap. As the handle of the auger is turned, downward pressure is applied on the crank. This turning and pushing motion works the auger through the trap and clears the blockage normally.

Flat-Tape Snakes

Flat-tape snakes are a perennial standby. You won't find many plumbers, regardless of how much high-tech drain-cleaning equipment they have, who fail to keep a flat-tape snake on their trucks. These snakes are available in various lengths and widths, and there are different heads for them. Before electric drain cleaners existed, these snakes were the cornerstone of drain cleaning.

Average lengths of flat-tape snakes are 25, 50, and 75 ft. The tape is coiled on a metal holder and is fished into a clogged drain as needed. Tape snakes work well in pipes with large diameters and few tight turns. However, they are not so effective in small drains or drains with short-turn fittings, especially galvanized-steel drains with elbows.

The principle behind these snakes is simple. You either break up the clog or push it through to a larger pipe. It is possible for any snake to become lodged in a tight turn or hard clog, so use caution when you force the snake into a drain.

Handheld Manual Spring Snakes

Small, handheld, manual-type spring snakes are sold in just about every store that caters to do-it-yourselfers. These little snakes can clear a tub or lavatory drain, but they are generally ineffective on anything else. The do-it-yourself versions are usually not of good quality and don't last long. Professional versions of these spinner-type snakes are available for plumbers on tight budgets, but it is better to save your money for an electric drain cleaner.

Handheld Electric Drain Cleaners

Handheld electric drain cleaners resemble hand drills with a large drum attached to them. These tools are great for sink, shower, tub, and lavatory drains. The spring cable can be fitted with different types of heads and can be made to accommodate the toughest of turns.

Electric shocks are possible if the insulation on the wire or housing of these units is damaged. When too much cable is left hanging out of a drain, the cable can flop wildly, splattering you with black gunk from the drain and causing uncomfortable bumps and bruises. Used properly, these machines are excellent for drains up to 3 in in diameter. Some models sit on a stand, instead of being handheld. Both types of machines work well, but most plumbers prefer the pistol-grip versions.

Big Drain Cleaners

Big drain cleaners mean big bucks. You spend big bucks to get them, and you make big bucks when you use them. These drain cleaners are used primarily for drains up to 4 in in diameter and come in many shapes and sizes. They are powerful and can hurt you badly. Most models are mounted on a frame that allows the unit to be rolled on wheels to the location.

Some models have removable drums that store the large

spring cables, and on other models the cable is fed through the machine from independent rolls. I prefer the type with the removable drum. Many models are equipped with a foot control that allows the machine to be operated with the touch of a toe.

These midsize sewer machines are available with numerous options. The price for this type of equipment starts at $1000 and goes up. It is very important to follow the proper safety procedures when you use these powerful drain cleaners.

Supersize Sewer Machines

Supersize sewer machines are used for very large drains. If you do commercial work, you may need this type of machine. When the drain you have to clear has a diameter of 6 in or more, this is the piece of equipment you want.

Like other electric drain cleaners, these machines demand respect. You can easily lose a finger to the coils of a supersize sewer machine. If you have to fight clogs in big pipes or tree roots growing in a sewer, these big bruisers will get the job done.

Drain cleaning can become your entire business if you live in an area where there are enough people to support your specialty. Aside from having to be available on a 24-hour basis, drain cleaning is a very good aspect of the plumbing trade to get involved with. It's fairly simple and can be extremely profitable. Whether you clean drains as a supplement to other work or specialize in busting clogs, the work is not too difficult, and the money is good.

12

Materials Selection

To do plumbing in the best and most cost-effective manner, you must select the proper materials. Today's plumbers have so many materials to choose from that the decision of which material is best suited to the job can be perplexing. Take the water distribution system as an example. When you are trying to decide what type of pipe to run your potable water through, you will have many options available. You could choose polybutylene, chlorinated polyvinyl chloride (CPVC), or copper, just to name a few. If you decide to use copper, you must determine which copper to use. Will you use type M, type L, or type K? Can you use polyethylene or polyvinyl chloride (PVC)? As you can see, it can be confusing.

This chapter gives a tour of the materials approved for use and which uses they are approved for. A material approved for aboveground use may not be allowed below ground. A pipe suitable for cold water may not work with hot water. This chapter explains the different materials and offers suggestions about which materials are best suited for specific uses. In addition, you will learn about approved connections.

Choosing the proper materials will help you in several ways. By working with only approved materials, you won't have to do the same job twice. Failure to use approved materials can result in code officers requiring you to rip out work already installed, so that it can be replaced with proper materials. Effective material selection can save you money or make your bid for a job more competitive. By assessing all circumstances surrounding a job, your material selection can help you avoid problems later with callbacks and warranty work.

There are three primary plumbing codes in use. One code controls most of the southern states. Another code rules much of the west and some of the central states. A third code covers the east coast, New England, and west to a point where it meets the code used in most western locations. For the sake of simplicity, I have grouped the states affected by these codes into three zones. Refer to Figs. 16.1, 16.2, and 16.3 for a breakdown of the states in which the code is used. This information will help you as we discuss code requirements in the remainder of this chapter.

What Is an Approved Material?

An approved material is one designated for specific uses, as determined by the local code enforcement office. Many approved materials and their approved uses are described in plumbing code books. But not all approved materials are listed. The materials detailed in code books represent the most commonly encountered approved materials. However, in certain cases other materials may be approved for use. This is frequently true when a material exceeds the requirements listed in a code book.

The standards set in code books are normally the minimum acceptable standards. A product that exceeds these minimums values may be allowed but not mentioned in the code. For general use, the materials listed in code books will be sufficient for your needs. Approved materials should be marked in a manner to be easily identified. Normally this identification will take the form of an embossed or molded marking or an indelible marking.

The type of identification marking used is frequently

determined by the type of material. Brass and copper fittings are often stamped. Plastic fittings usually have a molded marking. Pipe typically carries a colored stripe and indelible letters. For example, type M copper has an indelible red stripe, type L copper has a blue stripe, and type K copper has a green stripe. A yellow stripe indicates a drain-waste-and-vent (DWV) copper. By glancing at these color codes, a code enforcement officer can quickly identify the type of copper being worked with.

A material is subject to local approval. For example, the water quality in a certain location may have an adverse effect on a particular type of pipe. In such a case, the local code enforcement office may deny the use of a material identified as an approved material in the code book. A similar deviation from the code may be made for pipe used below grade. The soils in some areas are not compatible with certain materials. Adjustments for local conditions may be made by the local authorities and must be checked to ensure the legality of material usage.

Materials approved for carrying potable water must not have a lead content of more than 8 percent. This applies to pipe and fittings. Solder used to join pipe and fittings for potable water use may not contain more than 0.2 percent lead. There are several brands of lead-free solder available. The 50:50 solder, once common to the industry, is no longer approved for potable water systems.

When materials are approved, it is often for specific uses. It is not enough to say that polyethylene pipe is approved for potable water use. In a sense, this is a true statement, but it has exceptions. Polyethylene pipe may not be used to convey hot water. Polyethylene is not rated to convey hot water. Hot water is defined as having a temperature of at least 110°F. However, polyethylene is approved to carry potable water as a water service pipe. So, you see, polyethylene is approved for potable water use, but its range of use is limited. This type of regulation confuses many people, but this chapter will clear up the confusion. Now, let's look at approved materials and their allowed uses.

Water Service Pipe

A water service pipe is a pipe extending from a potable water source, a well, or a municipal water main to the interior of a building. Once inside the building, the water service pipe becomes a water distribution pipe. This distinction is important, especially in zone 3. In zone 3, water service pipe may not extend more than 5 ft into a building, unless it is of a material approved for water distribution. For example, a polyethylene water service pipe would have to be converted to an approved water distribution material once it was inside the building. But a polybutylene water service pipe would not have to be converted, because polybutylene is an approved water distribution pipe.

A water service pipe must be rated for a pressure compatible with the pressure produced from the water source. Typically, a water service pipe should be rated for 160 pounds per square inch (psi) at a temperature of 73.4°F. This rating will not always be applicable. If the pressure at the water source is higher than 160 psi, the pipe's rating must be higher. On the other hand, if the water source is a well and the water pressure from the well is below 160 psi, the rating of the pipe may be allowed to be lower. It is a good working principle to use pipe with a minimum working pressure of 160 psi, even if a lower rating is approved. This allows for future changes that may increase the water pressure. The pipe must be rated to withstand the highest pressure developed from the water source.

Water Service Materials

Now that you know what a water service is, we are going to look at the materials approved for use as a water service. The materials are discussed in alphabetical order. Not all the materials listed are commonly used, but they are all approved materials. Remember, plumbing codes change, and local jurisdictions have the authority to alter the code to meet local standards. Before you use these materials, or any other information in this book, consult your local code enforcement office to confirm the present status of local code requirements.

Acrylonitrile butadiene styrene (ABS)

ABS pipe is a plastic pipe. It is normally used for drains and vents, but if properly rated, it can be used as a water-service pipe. It must meet certain specifications for pressure-rated potable water use and must be approved by the local authorities. Since ABS is almost never used as a water service pipe, it will be difficult to locate this material in a rating approved for water service applications.

Asbestos cement pipe

Asbestos cement pipe has been used for municipal water mains in the past, but it is not used much today. It is, however, still an approved material.

Brass pipe

Brass pipe is an approved material for water service piping, but it is rarely used. The complications of placing this metallic, threaded pipe below ground discourage its use. Brass pipe can also be used as a water distribution pipe.

Cast-iron pipe

Cast-iron pipe is approved for use as a water service pipe, but it would not be used for individual water supplies. Sometimes called *ductile* pipe, this pipe is used in large water mains.

Chlorinated polyvinyl chloride

CPVC is a white or cream-colored plastic pipe allowed for use in water services, when it is rated for potable water service. This pipe is not commonly used for water service, but it could be. CPVC is fairly fragile, especially when cold, and it requires joints if the run of pipe exceeds 20 ft. It is advisable to avoid joints in underground water services. CPVC can also be used as a water distribution pipe.

Copper pipe

Copper tubing is often referred to as copper pipe, but there is a difference. Copper pipe can be found with or without threads. Copper pipe is marked with a gray color code. Copper pipe is approved for water service use, but copper tubing is the copper most often used by plumbers.

Copper tubing

Copper tubing is the copper most plumbers use. It can be purchased as soft copper, in rolls, or as rigid copper, in lengths resembling pipe. Copper tubing, frequently called copper pipe in the trade, has long been used as the plumber's workhorse. It is approved for water service use and comes in many different grades.

The three types of copper normally used are types M, L, and K. The type refers to the wall thickness of the copper. Type K is the thickest, and type M is the thinnest, with type L in the middle. Type L is generally considered the most logical choice for a water service. It is thicker than type M and is less expensive than type K. All three types are approved for water service applications.

When used as a water service pipe, soft, rolled copper is employed. Using the coiled copper, plumbers can roll the tubing into the trench in a solid length, without joints. This reduces the risk of leaks at a later date. Copper water services are not as common as they once were. Polyethylene and polybutylene are quickly reducing the use of copper. The reason is twofold: The plastic pipes are less expensive and are generally less affected by corrosion and other soil-related problems. Copper can also be used as a water distribution pipe.

Galvanized-steel pipe

Galvanized-steel pipe is an approved material for water services, but it is not a good choice. The pipe is joined with threaded fittings. Over time, this pipe rusts. It can rust at the threaded areas, where the pipe walls are weakened, or inside the pipe. When the threaded areas rust, they can leak. When the interior of the pipe rusts, it can restrict the flow of water and reduce water pressure. Although this gray, metal pipe is still available, it is rarely used in modern applications. If desired, galvanized-steel pipe can be used as a water distribution pipe.

Polybutylene

Polybutylene (PB) may very well be the pipe of the future. It is an amazing product. Polybutylene is available in rolled coils and in straight lengths. Polybutylene received mixed reviews when it was introduced to the plumbing trade, but it is gaining popularity quickly. The problems experienced in the early years were not with the pipe, but with the connections made on the pipe. These problems have been corrected, and PB is showing up in more plumbing trucks every day.

Polybutylene is approved as a water service pipe. It is available as a water service only pipe and as a water service water distribution pipe. If the pipe is blue, it is intended only for water service use. If it is gray, it can be used for water service or water distribution. Before PB pipe can be used for potable water, it must be tested by a recognized testing agency and approved by local authorities. Polybutylene is relatively inexpensive, very flexible, and a good choice for most water service applications.

Polyethylene

Polyethylene (PE) is a black, or sometimes bluish, plastic pipe that is frequently used for water services. It resists chemical reactions, as does polybutylene, and it is fairly flexible. This pipe is available in long coils, allowing it to be rolled out for great distances without joints.

This pipe may be one of the most common materials used for water services. However, it is not rated as a water distribution pipe. PE pipe is subject to crimping in tight turns, but is a good pipe that will give years of satisfactory service.

Polyvinyl chloride pipe

PVC pipe is well known as a drain-and-vent pipe, but the PVC used for water services is not the same pipe. Both pipes are white, but the PVC used for water services must be rated for use with potable water.

PVC water pipe is not acceptable as a water distribution pipe. It is not approved for hot water usage. Remember, when we talk of water distribution pipes, we assume the building is supplied with hot and cold water. If the building's only water distribution is cold water, some pipes such as PVC and PE pipes can be used.

Water Distribution Pipe

What is water distribution pipe? Water distribution pipe is the piping located inside a building that delivers potable water to plumbing fixtures. The water distribution system is normally comprised of both hot and cold water. Because of this fact, fewer materials are approved for water distribution systems than for water service piping.

A determining factor in choosing a pipe for water distribution is the pipe's ability to handle hot water. A water distribution pipe must be approved for conveying hot water. In zone 2, this means a pipe rated for a minimum working pressure of 100 psi at a temperature of 180°F. In zone 3 the rating must be 80 psi at 180°F. The reason that the pressure rating is lower than that of a water service pipe is simple. If the water pressure coming into a building exceeds 80 psi, a pressure-reducing valve must be installed at the water service, to reduce the pressure to no more than 80 psi.

Water Distribution Materials

Many water distribution materials are also approved for use as water service materials. However, the reverse is not true. Not all water service materials are acceptable as water distribution materials. Again, in alphabetical order, we discuss the types of piping approved for water distribution.

Brass pipe

Brass pipe is suitable for water distribution, but it is not normally used in modern applications. Although once popular, brass pipe has been replaced, in preference, by many new types of materials. The new materials are easier to work with and usually provide longer service, with fewer problems.

Copper pipe and copper tubing

Both copper pipe and copper tubing are acceptable choices for water distribution. Copper tubing, sometimes mistakenly called pipe, is by far the more common choice. Copper pipe has been around for many years and has proved itself to be a good water distribution pipe. If water has an unusually high acidic content, copper can be subject to corrosion and pinhole leaks. If acidic water is suspected, a thick-wall copper or a plastic-type of pipe should be considered instead of a thin-walled copper.

In zones 2 and 3, types M, L, and K can be used above and below ground. But in zone 1 type M copper is not permitted when installed underground within a building.

Galvanized-steel pipe

Galvanized-steel pipe remains in the approved category, but it is hardly ever used in new plumbing systems. The characteristics of galvanized-steel pipe remove it from the competition. It is difficult to work with and is subject to rust-related problems. The rust can cause leaks as well as reduced water pressure and volume.

Polybutylene

Polybutylene is edging copper out of the picture in many locations. With its ease of installation, resistance to splitting during freezing conditions, and its low cost, polybutylene is a strong competitor. Add to this the fact that PB resists chemical reactions, and you have yet another advantage over copper. Many old-school plumbers are dubious of the gray, plastic pipe, but it is making a good reputation for itself.

Drain-Waste-and-Vent Pipe

Pipe used for the drain-waste-and-vent (DWV) system is considerably different from its water pipe cousins. The most noticeable difference is size. DWV pipes typically range from $1\frac{1}{2}$ to 4 in in diameter; some are smaller and others are much larger. When you first glance at the names of DWV pipe materials, they may seem the same as the water service variety, but there are differences. Remember, for plastic water pipes to carry potable water, they must be tested and approved for potable water use.

DWV materials

There are a number of materials approved for DWV purposes, but in practice, only a few are commonly found in modern plumbing applications. Let's look, in alphabetical order, at the materials approved for DWV systems (Figs. 12.1 and 12.2)

Acrylonitrile butadiene styrene. ABS pipe is black, or sometimes a dark gray color, and is labeled as a DWV pipe, when it is meant for DWV purposes. ABS is normally used

Figure 12.1 Materials approved for above-ground vents in zone 2.

Cast iron

ABS

PVC

Copper

Galvanized steel

Aluminum

Borosilicate glass

Brass

Figure 12.2 Materials approved for above-ground vents in zone 3.

Cast iron

ABS

PVC

Copper

Galvanized steel

Lead

Aluminum

Brass

as a DWV pipe, instead of water pipe. The standard weight rating for common DWV pipe is schedule 40.

ABS pipe is easy to work with, and it may be used above or below ground (Figs. 12.3, 12.4, and 12.5). It cuts well with a hacksaw or regular hand saw. This material is joined with a solvent-weld cement and rarely leaks, even in less than desirable installation circumstances. ABS was very popular for a long time, but in many areas it is being pushed aside by PVC pipe, the white plastic DWV pipe. ABS is extremely durable and can take hard abuse, without breaking or cracking.

Figure 12.3 Materials approved for underground vents in zone 1.

Cast iron
ABS*
PVC*
Copper
Brass
Lead

Note: These materials may not be used with buildings having more than three floors above grade.

Figure 12.4 Materials approved for underground vents in zone 2.

Cast iron
ABS
PVC
Copper
Aluminum
Borosilicate glass

Figure 12.5 Materials approved for underground vents in zone 3.

Cast iron
ABS
PVC
Copper

In zone 1, the use of ABS is restricted to certain types of structures. ABS may not be used in buildings that have more than three habitable floors. The building may have a buried basement, where at least one-half of the exterior wall sections are at ground level or below. The basement may not be used as habitable space. So it is possible to use

ABS in a four-story building, as long as the first-story pipe is buried in the ground, as stated above, and not used in living space.

Aluminum tubing.   Aluminum tubing is approved for above-ground use only. Aluminum tubing is not allowed in zone 1. Aluminum tubing is usually joined with mechanical joints and coated to prevent corrosive action. This material is, like most others, available in many sizes. The use of aluminum tubing has not become common for average plumbing installations in most regions.

Borosilicate glass.   Borosilicate glass pipe may be used above or below ground for DWV purposes in zone 2. Underground use requires a heavy schedule of pipe. In zones 1 and 3 this pipe is not an approved material. However, as with all regulations, local authorities have the power to amend them to suit local requirements.

Brass pipe.   Brass pipe could be used as a DWV pipe, but it rarely is. The degree of difficulty in working with it is one reason. In zones 2 and 3, brass pipe is not to be used below grade for DWV purposes.

Cast-iron pipe.   Cast-iron pipe has long been a favored DWV pipe. Cast iron has been used for many years and provides good service, for extended periods. The pipe is available in a hub-and-spigot style, the type used years ago, and in a hubless version. The hubless version is newer and is joined with mechanical joints (Fig. 12.6), resembling a rubber coupling, surrounded and compressed by a stainless-steel band.

The older, bell-and-spigot or hub-and-spigot type of cast iron is the type most often encountered during remodeling jobs. This type of cast iron was normally joined with the use of oakum and molten lead. However, today rubber adapters are available for creating joints with this type of pipe. These rubber adapters also allow plastic pipe to be mated to cast iron.

Cast-iron pipe is frequently referred to as *soil pipe*. This

Figure 12.6 Hubless cast-iron connector band.

nickname separates DWV cast iron from cast iron designed for use as potable water pipe. Cast iron is available as a service-weight pipe and as an extraheavy pipe. Service-weight cast iron is the most commonly used. Even though the cost of labor and materials for installing cast iron is more than for schedule 40 plastic, cast iron still sees frequent use, both above and below grade.

Cast iron is sometimes used in multifamily dwellings and custom homes to deaden the sound of drainage as it passes down the pipe in walls adjacent to living space. If there are chemical or heat concerns, cast iron is often chosen over plastic pipe.

Copper. Copper pipe is made in a DWV rating. This pipe is thin-walled and is identified by a yellow marking. The pipe is a good DWV pipe, but it is expensive and time-consuming to install. DWV copper is not normally used in new installations, unless extreme temperatures, such as those from a commercial dishwasher, warrant the use of a nonplastic pipe. DWV copper is approved for use above and below ground in zones 1 and 2. In zone 3 a minimum copper rating of type L is required for copper used underground for DWV purposes.

Galvanized-steel pipe. Galvanized-steel pipe keeps popping up as an approved material, but it is no longer a good choice for most plumbing jobs. As galvanized-steel pipe ages and rusts, the rough surface due to the rust is prone to catch debris and create blockages. Another disadvantage of galvanized-steel DWV pipe is the time it takes to install the material.

Galvanized-steel pipe is not allowed for underground use in DWV systems. When used for DWV purposes, galvanized pipe should not be installed closer than 6 in to the earth.

Lead pipe. Lead pipe is still an approved material, but like galvanized-steel pipe, it has little place in modern plumbing applications. In zone 2 lead pipe is not permitted for DWV installations. In zone 3 the use of lead pipe is limited to above-grade installations.

Polyvinyl chloride pipe. PVC is probably the leader in today's DWV pipe. This plastic pipe is white and is normally used in a rating of schedule 40. PVC pipe uses a solvent-weld joint, and it should be cleaned and primed before it is glued together. This pipe becomes brittle in cold weather. If PVC is dropped on a hard surface while the pipe is cold, it is likely to crack or shatter. The cracks can go unnoticed until the pipe is installed and tested. Finding a cracked pipe, moments before an inspector is to arrive, is no fun. So be advised: Handle cold PVC with care. PVC may be used above or below ground.

In zone 1, the use of PVC is restricted to certain types of structures. PVC may not be used in buildings that have more than three habitable floors. The building may have a buried basement, where at least one-half of the exterior wall sections are at ground level or below. The basement may not be used as habitable space. So it is possible to use PVC in a four-story building, as long as the first-story pipe is buried in the ground, as stated above, and not used in living space.

If the underground piping is to be used as a building sewer, the following types of pipes may be used:

ABS	PVC
Cast iron	Concrete
Vitrified clay	Asbestos cement

In zone 3, bituminized-fiber pipe, type L copper pipe, and type K copper pipe may be used. Zone 1 tends to stick to the general guidelines, as given in previous paragraphs.

If a building sewer will be installed in a trench that contains a water service, some of the above pipes may not be used. Standard procedure for pipe selection under these conditions calls for the use of a pipe approved for use inside a building. This could include ABS, PVC, and cast iron. Pipes that are more likely to break, such as a clay pipe, are not allowed, unless special installation precautions are taken.

Sewers installed in unstable ground are also subject to modified rulings. Normally, any pipe approved for use underground inside a building will be approved for use with unstable ground. But the pipe must be well supported for its entire length.

Chemical wastes must be conveyed and vented with a system separate from the building's normal DWV system. The material requirements for chemical waste piping must be obtained from the local code enforcement office.

Storm Water Drain Materials

The materials used for interior and underground storm drainage (Figs. 12.7, 12.8, and 12.9) may generally be the same materials used for sanitary drainage. Any approved DWV material is normally approved for use in storm drainage.

Inside storm drainage

Materials commonly approved in zone 3 for interior storm drainage include the following:

Figure 12.7 Approved materials for storm water drainage in zone 2.

Materials approved for building storm water sewers
Cast iron
Aluminum*
ABS
PVC
Vitrified clay
Concrete
Asbestos cement

*Note: Buried aluminum must be coated.

Figure 12.8 Approved materials for storm water drainage in zone 2.

Materials approved for aboveground use
Galvanized steel
Black steel
Brass
DWV copper or thicker types of copper
Cast iron
ABS
PVC
Aluminum
Lead

Figure 12.9   Approved materials for storm water drainage in zone 2.

Materials approved for underground use
Cast iron
Coated aluminum
ABS*
PVC*
Copper*
Concrete*
Asbestos cement*
Vitrified clay*

*Note: These materials may be allowed for use, subject to local code authorities.

ABS	PVC
Type DWV copper	Type M copper
Type L copper	Type K copper
Asbestos cement	Bituminized fiber
Cast iron	Concrete
Vitrified clay	Aluminum
Brass	Lead
Galvanized steel	

In zone 2, approved pipe materials for interior storm drainage include

ABS	PVC
Type DWV copper	Type M copper
Type L copper	Type K copper
Asbestos cement	Cast iron
Concrete	Aluminum
Galvanized steel	Black steel
Brass	Lead

Zone 1 follows its standard pipe approvals for storm water piping (Figs. 12.10, 12.11, and 12.12).

Figure 12.10 Approved materials for storm water drainage in zone 1.

Materials approved for use inside buildings, below ground
Service-weight cast iron
DWV copper
ABS
PVC
Extra-strength vitrified clay

Figure 12.11 Approved material for storm water drainage in zone 1.

Material approved for use on exterior of buildings
Sheet metal with a minimum gauge of 26

Figure 12.12 Approved materials for storm water drainage in zone 1.

Materials approved for use inside buildings, above ground
Galvanized steel
Wrought iron
Brass
Copper
Cast iron
ABS*
PVC*
Lead

*Note: ABS and PVC may not be used in buildings that have more than three floors above grade.

Storm drainage sewers are a little different. If you are installing a storm drainage sewer, use one of the following types of pipes:

Zone 1 Follow the basic guidelines for approved piping.

Zone 2

Cast iron   Asbestos cement
Vitrified clay   Concrete
ABS   PVC
Aluminum (coated to prevent corrosion)

Zone 3

Cast iron   Asbestos cement
Vitrified clay   Concrete
ABS   PVC
Bituminized fiber   Type L copper
Type K copper   Type M copper
Type DWV copper

Subsoil Drains

Subsoil drains are designed to collect and drain water entering the soil. They are frequently slotted pipes and could be any of the following materials:

Asbestos cement   Bituminized fiber
Vitrified clay   PVC
Cast iron   Polyethylene

In zone 2, bituminized-fiber pipe cannot be used as a subsoil drain. Cast iron or some plastics may not be allowed. As always, check with local authorities before you use any material.

Other Types of Materials

We have concluded our look at the various types of pipes approved for plumbing, but there are other types of materials to consider. Valves, fittings, and nipples all fall under the watchful eye of the code enforcement office. In addition to these, there are still others to be discussed.

Fittings

Fittings that are made of cast iron, copper, plastic, steel, and other types of iron are all approved for use in their proper place (Figs. 12.13, 12.14, and 12.15). Generally speaking, fittings either must be made of the same material as the pipe they are being used with or must be compatible with the pipe.

Valves

Valves have to meet some standards, but most of the decision about the use of valves will come from the local code enforcement office. Valves, like fittings, must be either of

Figure 12.13 Fittings approved for horizontal changes in zone 1.

45° wye

Combination wye and eighth bend

Note: Other fittings with similar sweeps may also be approved.

Figure 12.14 Fittings approved for horizontal-to-vertical changes in zone 1.

45° wye

60° wye

Combination wye and eighth bend

Sanitary tee

Sanitary tapped-tee branches

Note: Cross fittings, like double sanitary tees, cannot be used when they have a short-sweep pattern; however, double sanitary tees can be used if the barrel of the tee is at least two pipe sizes larger than the largest inlet.

Figure 12.15 Fittings approved for vertical-to-horizontal changes in zone 1.

45° branches

60° branches and offsets, if they are installed in a true vertical position the same material as the pipe they are being used with or compatible with the pipe. Size and construction requirements are stipulated by local jurisdictions.

Nipples

Manufactured pipe nipples are normally made of brass or steel. These nipples range in length from $\frac{1}{8}$ to 12 in. Nipples must meet certain standards, and they must meet code requirements before they can be sold as an approved material.

Flanges

Closet flanges made of plastic must be $\frac{1}{4}$ in thick. Brass flanges may be only $\frac{1}{8}$ in thick. Flanges intended for caulking must be $\frac{1}{4}$ in thick with a caulking depth of 2 in. The screws or bolts used to secure flanges to a floor must be brass. All flanges must be approved for use by local authorities.

In zone 2 the use of offset flanges is prohibited unless there is prior approval. In zone 3, hard-lead flanges must weigh at least 25 ounces (oz) and must be made of a lead alloy with no less than a 7.75 percent antimony, by weight. In zone 1, flanges must have a diameter of about 7 in. In zone 1, the combination of the flange and the pipe receiving it must provide about $1\frac{1}{2}$ in of space to accept the wax ring or sealing gasket.

Cleanout plugs

Cleanout plugs are made of plastic or brass. Brass plugs are used only with metallic fittings. Unless they create a hazard, cleanout plugs must have raised, square heads. If the plugs are located where a hazard from the raised head may exist, countersunk heads may be used. In zone 2, borosilicate glass plugs must be used with cleanouts installed on borosilicate pipe.

Fixtures

Plumbing fixtures are regulated and must have smooth surfaces. These surfaces must be impervious. All fixtures must be in good working order and may not contain any hidden surfaces that may foul or become contaminated.

The rest

Lead bends and traps are not used much anymore, but if you decide to use these items, check with your code officer for guidance. These units must have a wall thickness of at least $\frac{1}{8}$ in.

The days of using sheet lead and copper to form shower pans are all but gone, but the code still offers minimum standards for these materials. Lead shower pans must not be rated at less than 4 pounds per square foot (lb/ft^2). If you need to use a lead pipe flashing, it should be rated at a minimum of 3 lb/ft^2. Copper shower pans should weigh 12 oz/ft^2, and copper flashings should have a minimum of 8 oz/ft^2.

When nonmetallic material is used for a shower pan, it must meet minimum standards. The material must be marked to indicate its approved qualities. Normally, membrane-type material is required to have a minimum thickness of 0.040 in. If the material must be joined at seams, it must be joined in accordance with the manufacturer's recommended procedure. Paper-type shower pans are also allowed for use when they meet minimum construction requirements.

Soldering bushings, once used to adapt lead pipe to other materials, and caulking ferrules, used in the conversion of cast iron to other materials, are all but a thing of the past. Lead pipe is usually removed in today's plumbing, and rubber adapters are used in place of caulking ferrules. If you have a nostalgic interest in these old-school items, you can find standards for them in your code book.

Connecting Materials

You now know what materials you can use for various purposes; next, you are going to learn how to connect them. Connection methods are normally approved on a local level. But there are some basics, and that's what I am going to cover.

Compatibility and performance

The main considerations in a good connection are compatibility and performance. A connection must be able to endure the pressure exerted on it in a normal testing procedure for the plumbing. For DWV connections, this usually amounts to a pressure of 4 to 5 pounds per square inch (psi). On water pipes, the pressure for a test should be equal to the highest pressure expected to be placed on the system, once it is in use. In zone 2, a pressure of at least 25 psi greater than the highest working pressure is required.

Basic preparation

Before pipe is connected with fittings, the pipe should be properly prepared. This means cutting the pipe evenly and clearing it of any burrs or obstructions. If a connection will be made with a different material from the pipe, the connector must be compatible with the pipe. For example, a rubber coupling held in place with stainless-steel clamps can be used to join numerous types of pipe styles.

ABS and PVC

Plastic pipes should be joined with solvent cements designed for the specific type of pipe. The plastic pipe and fittings should be clean, dry, and grease-free, before a joint is made. When you work with plastic pipes, application of a cleaner and primer is often recommended prior to the application of solvent cement. Always follow the manufacturer's recommendations for joining pipe and fittings. For best results, once the pipe has been coated with cement and inserted in a fitting, the pipe should be turned about a quarter turn. This helps to spread the glue and makes a better joint.

ABS pipe and fittings tend to harden much more quickly than PVC. ABS is also less sensitive to dirt and water in its joints. This is not to say that you can ignore proper procedures with ABS, but it is more forgiving than PVC.

Unusual pipes

Some types of materials, such as asbestos cement pipe and bituminized-fiber pipe, are used only in mechanical joints. Refer to the manufacturer's recommendations, your code book, and the local code officer for the best methods of joining this type of pipe.

Cast-iron pipe

Cast-iron pipe can be joined to fittings by using hot lead, rubber doughnuts, rubber couplings, or special bands designed for use with hubless cast iron, just to name a few.

In the old days, cast iron was almost always joined with a caulked joint, using molten lead. This procedure is still used today. To make this kind of joint, oakum or hemp is placed into the hub of a cast-iron fitting, around the pipe being joined to the fitting. The oakum or hemp must be dry when it is installed, if it is to make a good joint. Once the packing is in place, molten lead is poured into the hub. A packing tool, basically a special chisel, is used with a hammer to drive the lead down deeper into the hub. Once the oakum becomes wet, it expands and seals the joint.

Without prior experience or the help of an instructor, this type of joint can be extremely dangerous to make. Hot lead can take the skin right off your bones, and if hot lead comes into contact with a wet surface, the lead can explode, causing personal injury. Proper clothing and safety gear are a necessity for this type of job, even for seasoned professionals.

If you are using lead to make joints in cast iron meant to carry potable water, you will not use oakum or hemp. Instead, you will use a rope packing. This type of packing has a higher density than oakum and is meant for use with potable water installations.

Rubber doughnuts, as they are called in the trade, offer an alternative to caulking with hot lead. These special rubber adapters are placed in the hub of a fitting and lubricated. The end of the pipe to be joined is inserted into the rubber gasket and is driven or pulled into place. There are special tools to join soil pipe in this manner, but some plumbers use a block of wood and a sledge hammer to drive the pipe into the doughnut.

If a mechanical joint is being used on cast iron meant for potable water, there must be an elastomeric gasket on the joint, held in place with an approved flange. One example of a mechanical joint with an elastomeric gasket and approved flange is the standard band used with hubless cast iron. But there are other types of mechanical joints available and approved for these types of joints.

Copper

When it comes to copper, the options for joints include compressions fittings, soldered joints, screw joints, and other types of mechanical joints. Solder used to join copper for potable water systems must contain less than 0.02 percent lead. Threaded joints must be sealed with an acceptable pipe compound or tape. Welded and brazed joints are two other ways to join copper. Flared joints are still another way to mate copper to its fittings. Unions are allowed for connecting copper and are usually required with the installation of water heaters.

CPVC pipe

CPVC pipe is the homeowner's friend. Homeowners flock to this plastic water pipe, because all they have to do is glue it together—or at least, that's what they think. CPVC is a finicky pipe. It must be primed with an approved primer before it is glued. If this step is ignored, the joint will not be as strong as it should be. CPVC also requires a long time for its joints to set up. This pipe must be clean, dry, and grease-free, before you begin the connection process. Don't eliminate the priming process; it is essential to a good joint.

Galvanized-steel pipe

Galvanized-steel pipe lends itself to threaded connections, but rubber couplings can also be used. Use the large rubber couplings, not the hubless cast-iron type. Remember to apply the appropriate pipe dope or tape to the threads if you are making a screw connection. There are other types of approved ways to connect galvanized-steel pipe, but these are the easiest.

Polybutylene

Polybutylene pipe is normally connected to insert fittings with special clamps. These clamps are installed with a crimping tool designed for use with PB pipe. Compression fittings are also allowed with PB pipe, but the ferrules should be nonmetallic, to avoid cuts in the pipe. Flaring is possible with a special tool, but it is rarely needed. Another method of joining PB pipe employs heat fusion. This process is not normally used with water distribution piping; crimp rings are the most common method of joining polybutylene pipe in a water distribution system.

Polyethylene

Polyethylene pipe is typically joined with insert fittings and stainless-steel clamps. The insert fitting is placed inside the pipe, and a stainless-steel clamp is applied outside the pipe, over the insert fitting's shank. For best results, use two clamps.

Fixtures

Fixtures must be connected to drains in an approved manner. For sinks and such, the typical connection is made with slip nuts and washers. The washers may be rubber or nylon. For toilets, the seal is usually made with a wax ring.

Pipe penetrations

Pipe penetrations must be sealed to protect against water infiltration, the spreading of fire, and rodent activity. When a pipe penetrates a wall or roof, it must be adequately sealed to prevent the above-mentioned problems. Many styles of roof flashings are available for sealing pipe holes.

This concludes the discussion of commonly approved materials and their methods of connection. Always refer to your local plumbing code before you make a firm commitment to a particular material. Even though I have described the three major codes for you, local code officials have the power to amend codes to suit their requirements. To be safe, you have to check the local code for current regulations.

13

Plumbing Appliances

There are not many appliances that plumbers must work with, but there are a few. In some cases the appliances may be connected to waste or water pipes only by plumbers. In other situations, such as those involving water heaters, plumbers must have a good working knowledge of the appliance or device. Dishwashers, garbage disposals, and washing machines are three major plumbing appliances most plumbers get involved with. Ice makers are a fourth.

In terms of plumbing connections, none of these appliances is very difficult to troubleshoot. However, when it comes to the inner parts of these same appliances, troubleshooting can become more difficult.

This chapter reviews all the appliances mentioned above and covers effective troubleshooting procedures for each. We start in the kitchen and work our way into the mechanical room.

Garbage Disposals

Garbage disposals are a common appliance in many homes. Although they rarely work on the motors of disposals, plumbers are often the first service people called when a disposal is acting up. Problems with disposals fall into three primary categories: plumbing, electrical, and appliance repair.

The primary purpose of this chapter is to distinguish the causes of particular problems. While you, as a plumber, may not perform electrical work, you may be required to advise the property owner to call an electrician. Plumbers must be able to correct plumbing problems and refer other types of problems to appropriate repair people. As an example of knowing when to refer a customer to another type of service provider, let's consider the following scenario.

A homeowner calls you to repair a garbage disposal. When you arrive, you find that the disposal cuts on, but almost immediately it trips the reset button and cuts off. After testing the disposal a few times, you are convinced the problem is most likely in the electric wiring. Unless you are a licensed electrician, you should give the homeowner your opinion and suggest that perhaps an electrician should be called.

Under these conditions, you have fulfilled your role as a plumber. You responded to the call and did the initial troubleshooting. Seeing that the problem was not plumbing-related, you gave your opinion and suggestions. That is about all that anyone could expect of you.

If you are installing a garbage disposal, it is attached to the drain of the sink bowl. The disposal eliminates the need for a basket strainer. The finished part of the disposal's drain is installed like that of a basket strainer. From below, you slide a flange over the drain. The flange is held in place with a snap ring that fits on the drain collar. Tighten threaded rods to secure the drain to the sink. When the drain is tight, the disposal is mounted on the drain collar.

The mounting is usually done by holding the disposal on the collar and turning it clockwise. There is a small, plastic elbow to install on the side of the disposal. You will see a hole in the side of the disposal with a metal ring attached to it by screws. Remove the metal ring, and slide it over the elbow until it reaches the flange on the elbow. Place the supplied rubber gasket on the face of the elbow. Hold the elbow in place, and attach the metal bracket to the side of the disposal with the supplied screws. The trap or continuous waste attaches to the elbow with slip nuts and washers.

Now then, what can go wrong with a garbage disposal? Well, quite a bit. To expand on this, let's take the probable troubleshooting phases one at a time. This takes the form of stories about actual service calls.

In our first story, you have been called to a home to fix a garbage disposal that will not work. You arrive and turn on the garbage disposal. It makes a loud whirring sound, but doesn't dispose of its contents. What do you think is wrong? If you said it's jammed, you're right. Now, how can you fix it?

Disposals that are suffering from jammed impeller blades can be fixed in one of three ways, depending on the type of disposal.

Expensive disposals often have a switch that allows the rotation of the impellers to be reversed, thereby freeing them from whatever is jamming them. This type of repair, when done by a plumber, often makes homeowners feel foolish, and many plumbers can't look customers in the eyes when presenting the bill. As embarrassing as this situation can be for both parties, it is often simply a matter of turning on the reversing switch.

Many models of disposals are not equipped with reversing switches. A good number of these lower-grade disposals are, however, equipped with a special socket that allows the flywheel to be rotated. These sockets are usually hexagonal and are located on the bottom of the unit. Wrenches for the socket are supplied with the disposals. A regular Allen wrench will get the job done if the factory-supplied wrench is not available. Once you have a wrench in the socket, turn the flywheel with the wrench to free the jam.

Another common method of freeing jammed impellers in disposals involves the use of a broom handle or similar tool. The broom handle is placed into the mouth of the disposal and lodged against one of the impeller blades. Then pressure is exerted on the blade as the handle is pried back to free the jam. The disposal should be turned off for all these procedures except the use of a reversing switch.

In our second service call, you are called to a house where a disposal will not work. You turn on the switch, and nothing happens. What do you do?

The first logical step is to press the reset button, located on the bottom of the disposal. Turn off the disposal and press the reset button. Then turn on the switch and see whether the disposal works. If it runs, test it several times to make sure it does not continue to trip the safety control.

What do you do if the disposal stills fails to operate after you have pressed the reset button? Check to see whether the impellers are jammed. For obvious reasons, never put your fingers or hands into the mouth of the disposal. Try turning the flywheel with a wrench or some long tool, such as a broom handle. After you rotate the flywheel, turn the disposal back on and see if it works.

If it is still dead, check the electric panel to see if a fuse has blown or a circuit breaker has been tripped. You'd be surprised how often this is the cause of the problem. A lot of homeowners, and commercial customers for that matter, never think to check their electric panels before calling for help.

If the power to the disposal is on and it still won't run, you have two options. You can tell the customer the problem is not a plumbing problem and suggest calling an electrician or appliance repair person. Or, if you want to dig deeper, remove the wiring and test it to see if power is coming into the appliance. If there is full power at the wires, the problem is likely to be with the disposal's motor.

Our third service call puts you in the position of troubleshooting foul odors coming from a disposal. The homeowner has used the disposal to gobble up everything from fish to fowl, and now the odors are permeating the house.

You arrive on the job and notice the smell as soon as you walk through the door. The odor is not that of sewer gas, but it certainly is unpleasant. What do you do first?

Are you thinking you should check the trap first? If so, you're on the right track. If the water seal for the trap is not tight, odors could be rising from the drainpipe. However, since the sink and disposal are being used regularly, it is unlikely the trap has lost its seal. Unless a vent has become obstructed and is allowing siphonic action at the trap, there should be a good water seal. To be safe, check the trap. It is full of water, so the smell is not coming from the drainpipe. At this point you could tell the homeowner that the odor is coming from the appliance, not the plumbing, and leave. This would be an acceptable course of action, but there is a better one, if you know what to say. Do you?

Garbage disposals do sometimes become too smelly for normal comfort. When this happens, there is a simple way to beat the smell. Fill the disposal about two-thirds full with ice cubes and turn it on. Run cold water into the disposal to flush it out. After the ice cubes have been ground up, slice a lemon in half and put it in the disposal. Cut on the disposal and allow it to devour the lemon. This home remedy works on most any type of tough odor coming from disposals.

If a homeowner asks you whether it is better to flush a disposal with cold water or hot, what should you say? A lot of people, plumbers included, would say hot water. In fact, I used to think hot water was better for flushing a disposal. I found out, however, that cold water should be used because it congeals grease and carries it down the drain better than hot water. There is a greater likelihood of grease being caught on the sides of drain pipes if hot water is used.

Pure plumbing problems with disposals

Let's take a few moments to look at the pure plumbing problems often encountered with disposals. One of the most common problems is water leaking under the drain flange in the sink bowl. This is caused by a lack of putty under the flange. If this type of leak occurs on the back side of a disposal, water from the leak can appear to come from somewhere else, especially from the trap. If water is collecting beneath a sink bowl that is equipped with a disposal, check the drain flange carefully.

The lock ring that holds a disposal in place can become loose, allowing water to leak. These occasions are rare, but they do sometimes happen.

Occasionally, the gasket between the disposal and the drain ell leaks. The leak could be a result of a bad washer or a loose connection. A few quick turns of the lock screw with a screwdriver will tell you if the problem is a loose connection or a bad washer.

It is also not uncommon for the slip nut that holds the discharge end of the drain ell in a trap to leak. This usually requires nothing more than tightening the slip nut, but a new slip-nut washer may be needed.

The Mystery Leak of a Disposal   I'd like to tell you a story about the mystery leak of a disposal that perplexed two plumbers in my employ before the problem was solved. This happened a few years ago, and it is the only time I've ever heard of such a problem arising. Since this is a rare and unusual occurrence, I feel it warrants explanation.

My company received a call from a homeowner who was complaining of water under the kitchen sink. A plumber was dispatched to the call, but he couldn't find the source of the leak. The plumber was licensed, but young and inexperienced. When he couldn't figure out the problem, a more seasoned plumber was dispatched.

The experienced plumber went through all the typical troubleshooting steps. After his best effort, he couldn't find the leak. He gave up and called me. I was out on the road and not too far away, so I went over to the house.

When I arrived, there was solid evidence that something was leaking into the base cabinet, under the sink. I conferred with my plumber and then began to troubleshoot the job.

I could see all the water connections, and none was even damp. I filled the left bowl of the sink with water and emptied it. Nothing leaked. Then I filled the right bowl, the one to which the disposal was attached. The top to the disposal was missing, so I stuffed a rag in the opening to allow me to fill the sink. This was one step neither of the plumbers sent earlier had taken.

When the right bowl was nearly full, I removed the rag and turned on the disposal. Guess what happened? The kitchen cabinet began flooding with water. The look on my plumber's face was worth the time I had spent on the call. He was somewhere between embarrassed and amazed. And I must admit, I was surprised to see what was happening. Do you know where the water was coming from?

The water was blowing out of the fitting where a dishwasher hose can be connected to a disposal. Apparently, someone had knocked out the drain plug in the side outlet of the disposal and never connected anything to it. Nor did that person cap it. When the disposal was cut on, water was forced out of the uncovered drain outlet.

We capped the hole in the side of the disposal, and the problem never recurred. My plumbers made two mistakes in their troubleshooting. First, they never turned on the disposal. Second, they didn't fill the bowl with the disposal to capacity and drain it. All they had done was run water down the drain from the faucet.

Even though I had no idea the water was coming out of the drain outlet, I took all the right troubleshooting steps. I was surprised at the outcome of my test, but I was successful in finding and fixing the problem.

What is the moral to this story? It is simple, really. Take all the proper steps, don't take anything for granted, and you can probably solve any problem.

Ice Makers

When you have an ice maker to hook up, use a self-piercing saddle valve and $\frac{1}{4}$-in tubing. The tubing connects at the back of the refrigerator with a $\frac{1}{4}$-in compression fitting. The tubing may run to the water supply at the kitchen sink or to another more accessible cold water pipe. The saddle valve clamps around the cold water pipe and is held in place by two bolts.

Before you bolt on the saddle, make sure the rubber gasket that came with the saddle is in place. It should be in the hollow of the saddle, where the pipe will be pierced. Secure the saddle and connect the tubing to it. This connection is made with a compression fitting. Turn the handle clockwise until you cannot turn it any farther. Then turn the handle counterclockwise to open the valve and allow water to run through the tubing.

When the ice maker fails to function properly, it is not unusual for a plumber to be called. From a pure plumbing point of view, the only aspect of an ice maker that should concern a plumber is the connection at the back of the refrigerator, the tubing, and the connection to the cold water pipe. These aspects of ice makers can be troublesome, but the problems often lie within the ice maker unit. When the problem is with the unit itself, the troubleshooting and repair are in someone else's field of expertise. This is not to say that some plumbers won't work on ice maker units, but the units are not specifically plumbing.

If you, as a plumber, go poking around with an ice maker unit without the knowledge and experience required to fix it, you could be buying yourself a lot of trouble. Suppose your dabbling winds up breaking the ice maker? Guess who is going to have to pay to have it repaired or replaced? So restrict your work to only those areas where you are an expert. If you are not familiar with the workings of ice makers, limit your work to the plumbing connections and tubing. There is nothing wrong with telling customers that their problem is not plumbing-related when the trouble is the appliance.

What types of plumbing problems do ice makers have? Not too many, really. The saddle valve can create problems, and the tubing can leak, but that's really about all. Let's begin our short study of ice makers with the saddle valve.

Saddle valves don't normally give plumbers much trouble. If the valve is not working properly, it is inexpensive enough to just replace. Probably the most common complaint about saddle valves is that they leak at the packing nut. This is a simple problem—and one that is easily corrected by tightening the packing nut.

Sometimes a saddle valve will leak where the saddle comes into contact with the water distribution pipe. Usually the mounting screws are too loose, or possibly the gasket material is bad. Either cause is easily corrected.

The connection where the ice maker tubing is attached to

the refrigerator is another common place for leaks. If the refrigerator is moved, say, for cleaning purposes, the compression joint can become loose and leak. Again, this usually requires nothing more than tightening the nut.

If the tubing becomes kinked, the water flow to the ice maker can be reduced or even cut off completely. This can also happen if the refrigerator is moved.

Plastic tubing sometimes gets cut when a refrigerator is being moved, and I've heard stories of plastic tubing just bursting for no apparent reason, though I've never seen that happen personally. Obviously, if the tubing ruptures or is cut, a serious leak will ensue.

When you are asked to troubleshoot an ice maker problem, first check the saddle valve. Make sure the valve is turned on. I don't know how these valves get cut off, but I've found several over the years that were in the closed position.

If the valve is open, pull the refrigerator out, so that you can check the tubing connection to the appliance. Be careful not to tear up the kitchen floor when you move the refrigerator.

Loosen the nut on the connection at the back of the refrigerator. If you begin to get a good spray of water, you know the appliance is receiving the water it needs to work. When this is the case, your job, as a plumber, is complete. You have proved that the appliance is being supplied with water, and that is all plumbers should be expected to do.

If for some reason there isn't any water pressure at the back of the appliance, backtrack to the saddle valve. Loosen the connection nut on the tubing at the saddle valve. If there is water at the valve and no water at the appliance, obviously there is a problem in the tubing, probably a kink or possibly some type of obstruction.

If there is no water at the connection between the tubing and the saddle valve, it is either a closed valve or a bad valve. Try opening the valve, and if that doesn't work, replace it.

Dishwashers

Dishwashers have $\frac{5}{8}$-in drain hoses attached to them. This drain enters the cabinet under the kitchen sink and is attached to an air gap. An air gap is a device that mounts on the sink or countertop and has a wye connection. To install the air gap, remove the chrome cover by pulling on it. Remove the mounting nut from the threads. Push the air gap up through a hole in the sink or the counter from below. Install the mounting nut, and replace the chrome cover.

There are two ridged connection points on the air gap. The first accepts the $\frac{5}{8}$-in hose from the dishwasher. Before you put the hose on the air gap, slide a stainless-steel clamp over the hose. Next, push the hose onto the ridged connection, and tighten the clamp around the hose and insert connection. Attach a piece of $\frac{7}{8}$-in hose to the other connection point, using the same procedure.

Where the $\frac{7}{8}$-in hose leaves the air gap, it attaches to a garbage disposal or a wye tailpiece. The wye tailpiece attaches to the sink's tailpiece, like a tailpiece extension. The hose attaches to the wye portion of the extended tail-piece and is held in place with a stainless-steel clamp. If you are connecting to a disposal, knock out the plug in the disposal first. A thin, metal disk seated in the side of the disposal must be removed. This is normally done with a hammer and a screwdriver. Once the plug is removed, attach the hose to the connection with a stainless-steel clamp.

There is a small box under the dishwasher, where the water supply is connected. Buy a dishwasher ell to make the connection. One end of this elbow has male threads, and the other end is a compression fitting. Apply pipe dope to the threads, and screw it into the dishwasher. Under the kitchen sink, you need to tap into the hot water.

The hot water connection can be made with a tee fitting or a special type of cutoff stop. The special stops are designed to feed the sink and a dishwasher. The connection point for the dishwasher tubing will be the correct size, without any type of adapter. If you install a tee fitting, the dishwasher must have its own cutoff valve. A stop-and-waste valve is the type normally used. If you use a $\frac{1}{2}$-in valve, you will need a reducing coupling. The tubing going to the dishwasher has an inside diameter of $\frac{3}{8}$ in. The reducing coupling is $\frac{1}{2}$ in by $\frac{3}{8}$ in. Connect the tubing to the reducing coupling and run it to the dishwasher elbow. Make the connection at the compression fitting, and you're done.

Dishwashers have become almost standard equipment in modern homes and commercial kitchens. Even though problems that often occur with dishwashers are frequently appliance-related, many people turn to plumbers for help. There are some problems associated with dishwashers that are pure plumbing problems, and many that aren't. We concentrate on the pure plumbing problems and touch on the appliance-related problems.

Dishwasher won't drain

If the dishwasher won't drain, what do you check first? There are several possibilities, so let's look at each individually.

Trap obstructions. Trap obstructions are one of the first things to consider when you troubleshoot a dishwasher that will not drain. This is an easy defect to check for. Fill the kitchen sink bowl that uses the same drain as the dishwasher, and release the water to see whether the sink drains properly. If it does, rule out the trap. If it doesn't, disassemble the trap and investigate.

Drainpipe obstructions. Drainpipe obstructions may be found when you test the trap serving a dishwasher. If the sink using the same drainpipe as that used by the appliance will not drain, you can count on the problem being in the drainage system. After inspecting the trap, you will know whether the problem is in the trap or the drainpipe. If the pipe is clogged, snake it to clear the stoppage.

Air-gap obstructions.   Air-gap obstructions are rare, but they can prevent a dishwasher from draining properly. This type of condition is usually noticed when water draining from the dishwasher spills out of the air gap. Running a piece of wire through the air gap should solve the problem.

Kinked drain hoses.   Kinked drain hoses are another common cause of poor drainage from dishwashers. To check for this, remove the access panel on the front of the appliance. Look to see whether the hose is crimped.

Plugged disposal drainage inlets.   Plugged disposal drainage inlets are a sure cause of the failure of water to drain from a dishwasher. Some rookie plumbers, and a lot of do-it-yourselfers, don't know that a plug must be knocked out of a disposal drain inlet prior to its connection to the dishwasher drain hose.

I've been to more than a few jobs where dishwashers were flooding kitchens through the air gap because the knockout plug in a disposal hadn't been removed. To experienced plumbers, it seems so silly that sometimes this goes unchecked, but check it anyway. Remember, never take anything for granted. This is one of those times—never assume the installer of the dishwasher did the job correctly.

Stopped-up strainer baskets.   Stopped-up strainer baskets, on the inside of dishwashers, can inhibit proper drainage. Although this is not a pure plumbing problem, it is one you should check. If the strainer is clogged with grease, food, or other debris, a quick cleaning will get the appliance back in operation.

Water to dishwasher won't cut off

What is likely to be the problem with a dishwasher in which the water won't cut off during filling? There are only two probable causes for this situation. Either the solenoid is bad, or the inlet valve is malfunctioning. Neither of these falls into the pure plumbing category, but you should be aware of them and their symptoms.

If you choose to get involved with appliance repair, you may have to repair or replace the solenoid. If the problem is with the inlet valve, disassemble it and clean it. You may even have to replace it.

No water

What's wrong when there is no water coming into a dishwasher? A common cause is a closed valve on the supply tubing. Make sure the supply valve is in the open position.

Assuming the valve is open, there may be a faulty solenoid, a blocked inlet screen, crimped tubing, or, on a longshot, inadequate water pressure.

If the problem is the solenoid, it will have to be repaired or replaced; but again, this is not pure plumbing work.

The screen in the water inlet valve may be clogged with mineral deposits. Check this and clean the screen, if necessary.

Low water pressure is so unlikely that it is hardly worth mentioning. But if you suspect the pressure in the plumbing system is too low, you can put a pressure gauge on a hose bib to test it. As long as it reads normal pressure, something in the range of 40 psi, it should be all right.

Crimped tubing is not a likely cause when you are working with a dishwasher that has worked properly in the past, unless it was recently moved and reinstalled. Drop the front access panel, and look at the tubing to be sure it is not crimped.

Dishwasher isn't cleaning properly

What should you look for with a dishwasher that isn't cleaning dishes properly? There are five typical things to check. First, the water temperature should be checked. If the water being used in the dishwasher is not hot, it will not clean as well as it should. The water should be as hot as code allows. When possible, a temperature of 160°F is desirable.

Second, hard water can make dishwashers appear to not be doing their jobs. The film and residue left by hard water often make dishes appear to have not been cleaned well. If you suspect this to be the problem, test the water for hardness.

Third, water pressure has an effect on how well a dishwasher works. If the water pressure to the appliance is too low, the spray arm cannot work as well, and the water will not beat the dishes clean as it is designed to. Low water pressure is rarely a problem, but it could be.

Fourth, check to see that the spray arm is not jammed. Twirl it around by hand, and make sure nothing is inhibiting its rotation.

Fifth, while you are working with the spray arm, check that the holes are not plugged up. If the holes are blocked, water cannot be distributed to clean the dishes properly. Open blocked holes with a piece of wire.

Leaks

Leaks from dishwashers have several origins. In addition to the expected plumbing leaks, water can leak past a faulty door seal. Inspect the door seal to see that it is not worn, torn, or out of position.

Water leaks under dishwashers can come from the compression fitting at the dishwasher ell or from the threaded portion of the ell. It is also possible that the supply tubing has a hole in it. These leaks can usually be found easily by removing the access panel and inspecting the areas with a flashlight.

Drainage leaks are also possible under dishwashers. If the hose clamps are loose, leaks are likely. In addition to loose clamps, the insert fitting on the discharge outlet of the dishwasher may become cracked or broken. This doesn't happen often, but it can occur.

Washing Machines

Washing machines don't present many troubleshooting challenges. There are, however, a few tips worth mentioning.

Leaks around the hose connections of washing machines are usually due to bad or missing hose washers. By turning off the water valves and removing the washing machine hoses, you can do a quick inspection of the washers. If you suspect they are leaking, replace them.

Have you ever been called out because a washing machine would not fill with water? If you haven't, you probably will at some time. The major cause of this problem is debris in the hose screens. The troubleshooting process is simple:

Cut off the water at the washing machine valves. Remove the hoses from the valves, and place a bucket under the valves. Turn on the valves, and check whether there is good water pressure. There probably is. This tells you that the problem is somewhere in the hoses or in the machine itself.

Remove the hoses from the back of the washing machine. There are cone-shaped screens in the inlet ports. Inspect the screens for debris. If the screens are blocked, clean them or replace them. It is not uncommon for the screens to become so clogged that water pressure in the washing machine is either very low or zero. This is a common problem that baffles many inexperienced plumbers.

Although drainage leaks are not frequently a problem with washing machines, they do occur. If the drain hose is leaking, check the clamps where the hose connects to the back of the machine. Put the machine in a drain cycle, and check that the hose is not cracked. Cracks in rubber hose can be very difficult to see unless they are under water pressure.

Keep an eye on the washing machine drain receptor as the machine drains. If the standpipe is too short, water may be spilling out the top of the indirect waste pipe. This seems to happen most frequently when large amounts of detergent are used in the washing cycle.

Unless you are skilled in appliance repair, limit your work to plumbing. If an electrical problem exists, instruct the customer to call an electrician. When an ice maker is getting water but won't release ice cubes, suggest that the customer call an ice maker technician. Use common sense, and don't get yourself into a situation that you are not qualified to deal with.

14

Basics of Electricity

Plumbers often need some knowledge of the basics of electricity. Electrical work is underestimated by some and overestimated by others. In logical terms, electric wiring is not hard to understand. However, working with electricity can be very dangerous. Unlike plumbing, where a mistake is only likely to get you wet, a mistake with electricity can be fatal. Electricity should not be feared, but it should be respected.

Color Codes

Color codes are common in wiring. For example, a black or red wire is usually a hot wire. A white wire should be a neutral wire, but don't count on it; these wires can be hot. Green wires and plain copper wires are typically ground wires. When you match these colored wires to the various screws in an electric connection, they should go something like this: Black wires connect to brass screws. Red wires connect to brass or chrome screws. White wires normally connect to chrome screws. Green wires and plain copper wires connect to green screws. This is a pretty simple code, but it is one that anyone working with electric wiring should be aware of.

Wire Nuts

Wire nuts are available in different sizes and colors. The color indicates the size. Wire nuts are plastic on the outside and have wire springs on the inside. When wires are inserted into the wire nut, the nut can be turned clockwise to secure the wires. It may be necessary to twist the wires together before you install the wire nut. It is important to use a wire nut of the proper size. Wire nuts should be installed so that there is no exposed wiring.

Box Selection

Electric box selection can be pretty important. Depending upon the type of installation you are roughing in, you will need the right box for the job. Let's look at some options for electric boxes.

Rectangular boxes.   Rectangular boxes (Fig. 14.1) are commonly used for switches, wall outlets, and wall-mounted lights. Rectangular boxes are generally 3 in by 2 in.

Octagonal boxes.   Octagonal boxes (Fig. 14.2) are most frequently used for ceiling lights. They also may be used as a junction box, to join numerous wires together. These boxes typically have sides 4 in long.

Round boxes.   Round boxes are used primarily for ceiling lights.

Square boxes.   Square boxes are most often used as junction boxes. The side of a square box is typically from 4 in to just over $4\frac{1}{2}$ in.

Box depth

The box depth varies from $1\frac{1}{4}$ in to a full $3\frac{1}{2}$ in. This depth plays an important role in determining how many connections the box may hold. For example, a 3-in by 2-in switch box with a depth of $2\frac{1}{2}$ in can accommodate six connections with a no. 14 wire. The same box, but with a depth of $3\frac{1}{2}$

Figure 14.1   Rectangular electric box.

Figure 14.2   Octagonal electric box.

in, could hold nine connections. Ground wires usually count as a connection.

Means of attachment

The means of attachment for different boxes will vary, but some ways are easier than others. One of the easiest, and most common, boxes used in modern jobs is the plastic box with nails already inserted. These plastic boxes are nailed directly to wall studs. The boxes are sold with nails already inserted in a nailing sleeve. Simply position the box and drive the nail.

Some metal boxes are set up with a nailing flange that resembles an ear. Position the box and drive nails through the flange into the stud. This type of box is good because the ear is preset to allow for the thickness of common drywall.

Other metal boxes, usually meant for use in existing construction, are equipped with adjustable tabs. These tabs, or *ears* as they are sometimes called, allow the box to be attached to plaster lathe or to any other material that will accept and hold screws.

Octagonal and round boxes are often nailed directly to ceiling joists. If these boxes need to be offset, such as in the middle of a joist bay, then metal bars can be used to support the boxes. The metal bars are adjustable and mount between ceiling joists or studs. Once the bar is in place, the box can be mounted to the bar.

Common Heights and Distances

There are some common heights and distances used in electric wiring. For example, wall switches are usually mounted about 4 ft above the finished floor. Wall outlets are normally set between 12 and 18 in above the floor. In most areas, wall outlets must be spaced so that there is not more than 12 ft between them.

Wire Molding

Wire molding is a protective trim that is placed over wires when the wires are run on the outside of walls and ceilings. This molding can be installed after wiring is run. The wire molding is hollow and open on the bottom; this allows the molding to be laid over existing wires. When you are unable to conceal wires via a conventional method, wire molding should be used to protect you and the wiring.

Conduit

Conduit is another option for running wires when conventional methods cannot be employed. However, wires must be pulled through the conduit. Conduit cannot be installed over existing wires. The conduit acts as a protective tunnel for the wires.

Snaking Wires

You snake wires through a wall with a fish tape. A fish tape is a thin, flexible, metal tape on a spool. The exposed end of the tape has a hook on it. Some people simply bend the wire to be fished over the hook on the fish tape. Seasoned electricians usually bend the wires over the hook and then tape them to ensure they don't come loose.

Fish tapes can be used to get wires up and down wall cavities. However, fire blocking in the walls can make a fish tape almost useless. It is not unusual for walls to have staggered blocks of wood nailed between the studs. These wood blocks help to block airflow and prevent the wall cavities from acting as chimneys in the event of a fire. If fire blocking is encountered, the wall must be opened.

Fish tapes are often worked from around outlet boxes. Removing the cover of an outlet box gives you access to the interior of the wall. For example, if you need a wire in an attic during a conversion, you could drill a hole in the wall plate. Then, from below, you could remove an outlet cover as the second opening for working the fish tape. Between these two openings, you could fish a wire up or down the wall, to either opening.

Putting Wire under the Screw

When you are putting wire under a screw, the wire should be bent to tighten as the screw is tightened (Fig. 14.3). If you hook the wire in the opposite direction, it may come loose. Always twist the hook in the end of the wire to match the clockwise tightening of the screw.

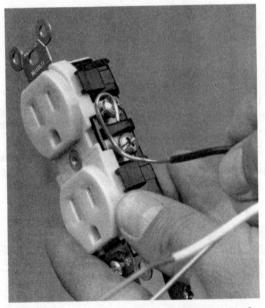

Figure 14.3 The proper way to bend electric wire for placement under a screw.

Ground Fault Interrupters

Ground fault interrupters (GFIs), sometimes called *ground fault interceptors, ground fault breakers,* or *ground fault outlets,* are usually required in any location where water may come into contact with the electric outlet. Bathrooms and kitchens in particular are generally required to be equipped with a GFI. Outdoor receptacles usually require a GFI.

There are two common ways to provide GFI protection. First, you can install a GFI outlet. These outlets have test-and-reset buttons. GFIs should be tested monthly. Second, install a GFI breaker. This is a special circuit breaker that is installed in the service panel. Either of these devices will protect against accidental shocks in a wet area. GFIs are not cheap, but in most areas they are a code requirement for wet-use areas, and they are an excellent safety feature.

High-Voltage Circuits

Appliances that use high voltage require their own high-voltage circuit. A clothes dryer or an electric range are in this category, and so are water heaters and well pumps. The outlets for these high-voltage circuits have different plug patterns from standard wall outlets. Appliances that use these special plugs are often fitted with pigtails. The pigtails are special wiring arrangements, with high-voltage prongs, that allow the appliance to be plugged into the high-voltage outlet. These circuits run directly from the outlet to the service panel.

Floor Outlets

Floor outlets are generally frowned upon by electricians and electric inspectors. When used, electric outlets mounted in the floor should be equipped with waterproof covers. When possible, avoid installing outlets in the floor.

As a plumber, your electric work should be limited. When a new installation is being made, such as a dishwasher or garbage disposal, the electric work should be done by a licensed electrician. However, it is not unusual for plumbers to do their own wiring when replacing existing dishwashers and disposals. Plumbers also frequently do their own wiring when replacing electric water heaters. So let's talk about these three types of jobs in a little greater depth.

Dishwashers

The wiring in a dishwasher connection is very simple. There are only three wires to work with, and one is a ground wire. The ground wire goes under a ground screw, and the other wires are matched up by color. Wire nuts are used to secure connections between the primary wires. There really isn't much involved in changing a dishwasher. However, make sure that all power to the circuit is off while the electric work is being done.

Dishwashers are supposed to be wired with a circuit that does not serve any other appliance or outlet. This was not always the case. Check the electric panel to determine which fuse or circuit breaker controls the dishwasher. Once you find it, cut it off. If you want to be really safe, turn on the dishwasher and let it run while you shut off the power. This will ensure that you have cut off the power properly. Even so, check the wires with your electric meter before you touch the bare ends. You can never be too careful when working with electricity.

Garbage Disposals

Garbage disposals are no more difficult to wire than dishwashers. The setup is basically the same. One difference, however, is the switch that turns the disposal on and off. Technically, you could work safely with the wiring as long as the switch is off. I don't trust this procedure, though. Cut off the power at the switch and at the panel box. This reduces the risk of someone turning on the switch by accident while you are working with the wires under the sink.

Water Heaters

Electric waters heaters should be equipped with a disconnect switch, but not all of them are. The voltage serving a water heater has potentially fatal power. Never assume anything with electric work. Cut off the disconnect switch, and flip the circuit breaker. Use an electric meter to confirm that the power is off before you touch the wires. Don't take any chances on getting shocked.

When it comes to electric work, remember that you are a plumber, not an electrician. I suggest that you leave as much electric work as you can for electricians to do. Most plumbers do take on some electric responsibilities in their routine service work. Don't do this until you understand electric work very well. Get comfortable with the test meter and use it. Remember, it only takes one mistake with electricity to ruin your day, your career, and your life.

15

Safety Procedures
and Precautions

Safety procedures and precautions don't get the attention
they deserve. Far too many people are injured on the job
every year. Most injuries could be prevented, but they are
not. Why? People are in a hurry to make a few extra bucks,
so they cut corners. This happens with plumbing contractors and piece workers. It even affects hourly plumbers who
want to shave 15 minutes off their workday, so that they
can head back to the shop early.

Based on my field experience, most accidents occur as a
result of negligence. Plumbers try to cut corners, and they
wind up getting hurt. This has been true for my personal
injuries. I've only suffered two serious on-the-job injuries,
and both were a direct result of my carelessness. I knew
better than to do what I was doing when I was hurt, but I
did it anyway. Well, sometimes you don't get a second
chance, and the life you affect may not be your own. So,
let's look at some sensible safety procedures that you can
implement in your daily activity.

Dangers in Plumbing

Plumbing can be a very dangerous trade. The tools of the
trade are potential killers. Requirements of the job can

place you in positions where a lack of concentration could result in serious injury or death. The fact that plumbing can be dangerous is no reason to rule out the trade as your profession. Driving can be extremely dangerous, but few people avoid automobiles because of fear.

Fear is generally a result of ignorance. When you have both knowledge and skill, fear begins to subside. As you become more accomplished at what you do, fear is forgotten. While it is advisable to learn to work without fear, you should never work without respect. There is a huge difference between fear and respect.

If you are afraid to climb up on a roof to flash a pipe, you are not going to last long in the plumbing trade. However, if you scurry up the roof recklessly, you could be injured severely, perhaps even killed. You must respect the position you are putting yourself in. If you are using a ladder to get onto the roof, you must respect the outcome of a possible mistake. Once you are on the roof, you must be conscious of your footing and how you negotiate the pitch of the roof. If you pay attention, are properly trained, and don't get careless, you are not likely to get hurt.

Being afraid of getting up on a roof will limit or eliminate your plumbing career, but treating your trip to and from the roof like a walk into your living room could be deadly. Respect is the key. If you respect the consequences of your actions, you are aware of what you are doing and the odds for a safe trip improve.

Many young plumbers are fearless in the beginning. They think nothing of darting around on a roof or jumping down into a sewer trench. As their careers progress, they usually hear of or see on-the-job accidents. Someone gets buried in the cave-in of a trench. Somebody falls off a roof. A metal ladder being set up hits a power line. A careless plumber steps into a flooded basement and is electrocuted because of submerged equipment. The list of possible job-related injuries is a long one.

Millions of people are hurt every year in job-related accidents. Most of these people were not following solid safety procedures. Sure, some were victims of unavoidable accidents, but most were hurt by their own hand, in one way or another. You don't have to be one of these statistics.

In nearly 20 years of plumbing, I have only been hurt seriously on the job twice. Both times were my fault. I got careless. In one instance, I let loose clothing and a powerful drill work together to chew up my arm. In the other, I tried to save myself the trouble of repositioning my stepladder while drilling holes in floor joists. My desire to save a minute cost me torn stomach muscles and months of pain caused by a twisting drill.

My accidents were not mistakes; I was stupid. Mistakes are made through ignorance. I wasn't ignorant of what could happen to me. I knew the risk I was taking, and I knew the proper way to perform my job. Even given that knowledge, I slipped up and got hurt. Luckily, both of my injuries healed, and I didn't pay a lifelong price for my stupidity.

During my long plumbing career I have seen a lot of people get hurt. Most of these people have been helpers and apprentices. Every one of the on-the-job accidents that I witnessed could have been avoided. Many of the incidents were not extremely serious, but a few were.

As a plumber, you will be doing some dangerous work. You will be drilling holes, running threading machines, snaking drains, installing roof flashings, and doing a lot of other potentially dangerous jobs. Hopefully, your employer will provide you with quality tools and equipment. If you have the right tool for the job, you are off to a good start in staying safe.

Safety training is something you should seek from your employer. Some plumbing contractors fail to tell employees how to do their jobs safely. It is easy for someone, such as an experienced plumber, who knows a job inside and out to forget to inform an inexperienced person of potential danger.

For example, a plumber might tell you to break up the concrete around a pipe, to allow the installation of a closet flange, and might never consider telling you to wear safety glasses. The plumber assumes you know the concrete is going to fly up in your face as it is chiseled up. However, being a rookie, you might not know about the reaction of concrete hit with a cold chisel. One swing of the hammer could cause extreme damage to your eye.

Simple jobs, like the one in the example, are all it takes to ruin a career. You might be really on your toes when asked to scoot across an I-beam, but how much thought are you going to give to tightening the bolts on a toilet? The risk of falling off the I-beam is obvious. Having chips of china, from a broken toilet where the nuts were turned one time too many, flying into your eyes is not so obvious. Either way, you can have a work-stopping injury.

Safety is a serious issue. Some job sites are very strict in their safety requirements. But on a lot of jobs there are no written rules of safety. If you are working on a commercial job, supervisors are likely to make sure you abide by the rules of the Occupational Safety and Health Administration (OSHA). Failure to comply with OSHA regulations can result in stiff financial penalties. However, if you are working in residential plumbing, you may never set foot on a job where OSHA regulations are observed.

In all cases, you are responsible for your own safety. Your employer and OSHA can help you to remain safe, but in the end it is up to you. You have to know what to do and how to do it. And not only do you have to take responsibility for your own actions, but also you have to watch out for the actions of others. You could be injured by someone else's carelessness. Now that you have had the primer course, let's get down to the specifics of job-related safety.

As we move into specifics, you will find the suggestions in this chapter broken down into various categories. Each category will deal with specific safety issues related to the category. For example, in the section on tool safety, you will learn procedures for working safely with tools. As you move from section to section, you may notice some overlapping of safety tips. For example, in the section on general safety, you will see that it is wise to work without wearing jewelry. Then you will find jewelry mentioned again in the tool section. The duplication is intended to pinpoint definite safety risks and procedures. We start with general safety.

1. Wear safety equipment.
2. Observe all safety rules at the particular location.
3. Be aware of any potential dangers in the specific situation.
4. Keep tools in good condition.

Figure 15.1   General safe working habits.

General Safety

General safety (Fig. 15.1) covers a lot of territory. It starts from the time you get into the company vehicle and carries you right through to the end of the day. Many of the general safety recommendations involve the use of common sense.

Vehicles

Many plumbers are given company trucks to use in getting to and from jobs. You will probably spend a lot of time loading and unloading company trucks. And, of course, you will spend time either riding in or driving them. All these activities can pose a threat to your safety.

If you will be driving the truck, take time to get used to how it handles. Loaded plumbing trucks don't drive the same as the family car. Remember to check the vehicle's fluids, tires, lights, and related equipment. Many plumbing trucks are old and have seen better days. Failure to check the vehicle's equipment could result in unwanted headaches. Also, remember to use the seat belts; they *do* save lives.

Apprentices are normally given the job of unloading the truck at the job site. There are a lot of ways to get hurt in doing this job. Many plumbing trucks use roof racks to haul pipe and ladders. If you are unloading these items, make sure they won't come into contact with low-hanging electric wires. Copper pipe and aluminum ladders make very good electric conductors, and they will carry the power surge through you on the way to the ground. If you are unloading heavy items, don't put your body in awkward positions. Learn the proper ways to lift, and never lift objects inappropriately. If it is wet, be careful when climbing on the truck. Step bumpers get slippery, and a fall can impale you on an object or bang up your knee.

When it is time to load the truck, observe the same safety precautions as in unloading. In addition, always make sure the load is packed evenly and is well secured. Be especially careful of any load that you attach to the roof rack, and always double-check the cargo doors on trucks with utility bodies.

Not only will it be embarrassing to lose your load while going down the road, it could be deadly. I once saw a one-piece fiberglass tub/shower unit fly out the back of a pickup truck as the truck was traveling on an interstate highway. As a young helper, I lost a load of pipe in the middle of a busy intersection. In that same year, the cargo doors on the utility body of my truck flew open as I came off a ramp, onto a major highway. Tools scattered across two lanes of traffic. These types of accidents don't have to happen. It's your job to make sure they don't.

Clothing

Clothing is responsible for a lot of on-the-job injuries. Sometimes it is the lack of clothing that causes the accidents, and there are many times when too much clothing creates the problem. Generally, it is wise not to wear loose-fitting clothes (Fig. 15.2). Shirt tails should be tucked in, and short-sleeve shirts are safer than long-sleeve shirts when you are operating some types of equipment.

Caps can save you from minor inconveniences, such as getting glue in your hair, and hard hats provide some protection from potentially damaging accidents, such as a steel fitting dropping on your head. If you have long hair, keep it up and under a hat.

Good footwear is essential in the trade. Normally a strong pair of hunting-style boots is best. The thick soles

1. Do not wear clothing that ignites easily.
2. Do not wear loose clothing, wide sleeves, ties, or jewelry (bracelets, necklaces) that can become caught in a tool or otherwise interfere with work. This caution is especially important when you are working with electric machinery.
3. Wear gloves to handle hot or cold pipes and fittings.
4. Wear heavy-duty boots. Avoid wearing sneakers on the job. Nails can easily penetrate sneakers and can cause a serious injury (especially if the nail is rusty).
5. Always tie shoelaces. Loose shoelaces can easily cause you to trip, possibly leading to injury to yourself or other workers.
6. Wear a hard hat on construction sites to protect your head from falling objects.

Figure 15.2 Safe dressing habits.

provide some protection from nails and other sharp objects that you may step on. Boots with steel toes can make a big difference in your physical well-being. If you are going to be climbing, wear foot gear with a flexible sole that grips well.

Gloves can keep your hands warm and clean, but they can also contribute to serious accidents. Wear gloves sparingly, depending on the job you are doing.

Jewelry

On the whole, jewelry should not be worn in the workplace. Rings can inflict deep cuts in your fingers. They can also work with machinery to amputate fingers. Chains and bracelets are equally dangerous—probably more so.

Protection of eyes and ears

Eye and ear protection is often overlooked. An inexpensive pair of safety glasses can prevent you from spending the rest of your life blind. Ear protection reduces the effect of loud noises, such as jackhammers and drills. You may not notice much benefit now, but in later years, you will be glad you wore it. If you don't want to lose your hearing, wear ear protection when you are subjected to loud noises.

Pads

Knee pads not only make a plumber's job more comfortable, but also help to protect the knees. Some plumbers spend a lot of time on their knees, and pads should be worn to ensure they can continue to work for many years.

The embarrassment factor plays a significant role in job-related injuries. People, especially young people, feel the need to fit in and to make a name for themselves. Plumbing is sort of a macho trade. It is no secret that plumbers often fancy themselves to be strong human specimens. Many plumbers are strong. The work can be hard and in doing it, becoming strong is a side benefit. But you can't allow safety to be pushed aside just to make a macho statement.

Too many people believe that working without safety glasses, ear protection, and so forth makes them look tough. That's just not true. It may make them look dumb, and it may get them hurt, but it does *not* make them look tough. If anything, it makes them look stupid or inexperienced.

Don't fall into that trap. Never let people goad you into bad safety practices. Some plumbers are going to laugh at your knee pads. Let them laugh—you will still have good knees when they are hobbling around on canes. I'm dead serious about this issue. There is nothing sissy about safety. Wear your gear in confidence, and don't let the few jokesters get to you.

Tool Safety

Tool safety (Fig. 15.3) is a big issue in plumbing. Plumbers work with numerous tools, all of which are potentially dangerous; but some are especially hazardous. This section discusses the various tools used on the job. You cannot afford to start working without knowing the basics of tool safety. The more you can absorb about tool safety, the better off you will be.

The best starting point is to read all literature available

1. Use the right tool for the job.
2. Read any instructions that come with the tool unless you are thoroughly familiar with its use.
3. Wipe and clean all tools after each use. If any other cleaning is necessary, do it periodically.
4. Keep tools in good condition. Chisels should be kept sharp, and any mushroomed heads kept ground smooth; saw blades should be kept sharp; pipe wrenches should be kept free of debris and the teeth kept clean; etc.
5. Do not carry small tools in your pocket, especially when you are working on a ladder or scaffolding. If you fall, the tools may penetrate your body and cause serious injury.

Figure 15.3 Safe use of hand tools.

from the manufacturers of your tools. The people who make the tools offer some good safety suggestions. Read and follow the manufacturers' recommendations.

The next step is to ask questions. If you don't understand how a tool operates, ask someone to explain it to you. Don't experiment on your own. The price you pay could be much too high.

Common sense is irreplaceable in the safe operation of tools. If you see an electric cord with cut insulation, you should have enough common sense to avoid using it. In addition to this type of simple observation, you will learn some interesting facts about tool safety. Now, let me tell you what I've learned about tool safety over the years.

There are some basic principles to apply to all your work with tools:

- Keep body parts away from moving parts.
- Don't work in poor lighting conditions.
- Be careful of wet areas when you are working with electric tools.
- If special clothing is recommended for working with your tools, wear it.

- Use tools only for their intended purposes.
- Get to know your tools well.
- Keep your tools in good condition.

Now, let's take a close look at the tools you are likely to use. Plumbers use a wide variety of hand tools and electric tools. They also use specialty tools. So let's see how you can use all these tools without injury.

Torches

Torches are often used by plumbers on a daily basis. Some plumbers use propane torches, and others use acetylene torches. In either case, the containers of fuel for these torches can be very dangerous. The flames produced from torches can also do a lot of damage.

When working with torches and tanks of fuel, you must be very careful (Fig. 15.4). Don't allow the torch equipment to fall on a hard surface—the valves may break. Check all connections closely. A leak allowing fuel to fill an area could result in an explosion when you light the torch.

Always pay attention to where the flame is pointed. Carelessness with the flame could start unwanted fires. This is especially true when you work near insulation and

---

1. Always keep fire extinguishers handy. Make sure that the extinguisher is full and that you know how to use it quickly.
2. Be sure to disconnect and bleed all hoses and regulators used in welding, brazing, soldering, etc.
3. Store cylinders of acetylene, propane, oxygen, and similar substances in an upright position in a well-vented area.
4. Operate all air-acetylene, welding, soldering, and related equipment according to the manufacturer's directions.
5. Do not use propane torches or other similar equipment near material that can easily catch fire.
6. Be careful at all times. Be prepared for the worst and be ready to act.

Figure 15.4 Preventing fires.

other flammable substances. If the flame is directed at concrete, the concrete may explode. Since moisture is retained in concrete, intense heat creates pressure on the moisture, forcing the concrete to explode.

It's not done often enough, but you should always have a fire extinguisher close by when working with a torch. If you have to work close to flammable substances, use a heat shield on the torch. When the flame is close to wood or insulation, try to remove the insulation or wet the flammable substance, before you use the flame. When you have finished using the torch, make sure the fuel tank is turned off and the hose is drained of all fuel. Use a striker to light your torch. Using a match or cigarette lighter puts your hand too close to the source of the flame.

Lead pots and ladles

Lead pots and ladles offer their own style of potential danger. Plumbers today don't use molten lead as much as they used to, but hot lead is still used. It is employed to make joints with cast-iron pipe and fittings.

When working with a small quantity of lead, many plumbers heat the lead in a ladle. They melt the lead with the torch and pour it straight from the ladle. When larger quantities of lead are needed, lead pots are used. The pots are filled with lead and set over a flame. All this type of work is dangerous.

Never put wet materials in hot lead. If the ladle is wet or cold when it is dipped into the pot of molten lead, the hot lead can explode. Don't add wet lead to the melting pot if it contains molten lead. Before you pour the hot lead into a waiting joint, make sure the joint is not wet. As another word of caution, don't leave a working lead pot where rain dripping off a roof can fall into it.

Obviously, molten lead is hot, so don't touch it. Be very careful to avoid tipping over the pot of hot lead. I remember one accident when a pot of hot lead was knocked over on a plumber's foot. The scene was terrifying.

Drills and bits

Drills have been my worst enemy in plumbing. The two serious injuries I suffered were both related to my work with a right-angle drill. The drills used most by plumbers are not the little pistol-grip, handheld types of drills most people think of. The day-to-day drilling done by plumbers involves the use of large, powerful right-angle drills. These drills have enormous power when they get in a bind. Hitting a nail or a knot in the wood being drilled can do a lot of damage. You can break fingers, lose teeth, suffer head injuries, and endure a lot more. As with all electric tools, always check the electric cord before you use the drill. If the cord is not in good shape, don't use the drill (Fig. 15.5).

Always know what you are drilling into. If you are doing new-construction work, it is fairly easy to look before you drill. However, drilling in a remodeling job can be much more difficult. You cannot always see what you are getting into. If you are unfortunate enough to drill into a hot wire, you can get a considerable electric shock.

The bits you use in a drill are part of the safe operation of the tool. If your drill bits are dull, sharpen them. Dull

1. Always use a three-prong plug with an electric tool.
2. Read all instructions concerning the use of the tool (unless you are thoroughly familiar with its use).
3. Make sure that all electrical equipment is properly grounded. Ground fault circuit interrupters (GFCIs) are required by OSHA regulations in many situations.
4. Use proper extension cords of a proper size. (Undersize wires can burn out a motor, cause damage to the equipment, and present a hazardous situation.)
5. Never run an extension cord through water or through any area where it can be cut, kinked, or run over by machinery.
6. Always hook up an extension cord to the equipment, and *then* plug it into the main electric outlet—not vice versa.
7. Coil up and store extension cords in a dry area.

Figure 15.5   The safe use of electric tools.

bits are much more dangerous than sharp ones. When you are using a standard plumber's bit to drill through thin wood, such as plywood, be careful. Once the worm driver of the bit penetrates the plywood fully, the big teeth on the bit can bite and jump, causing you to lose control of the drill. If you are drilling metal, be aware that the metal shavings are sharp and hot.

Power saws

The most common type of power saw used by plumbers is the reciprocating saw. These saws are used to cut pipe, plywood, floor joists, and a whole lot more. In addition to reciprocating saws, plumbers use circular saws and chop saws. All saws have the potential to cause serious injury.

Reciprocating saws are reasonably safe. Most models are insulated to help avoid electric shocks if a hot wire is cut. The blade is typically a safe distance from the user, and the saws are pretty easy to hold and control. However, the brittle blades do break. This could result in an eye injury.

Circular saws are used by plumbers occasionally. The blades on these saws can bind and cause the saws to kick back. Chop saws are sometimes used to cut pipe. If you keep your body parts out of the way and wear eye protection, chop saws are not unusually dangerous.

Handheld drain cleaners

Handheld drain cleaners don't see a lot of use by plumbers who do new-construction work, but they are a frequently used tool of service plumbers. Most of these drain cleaners resemble, to some extent, a straight, handheld drill. There are models that sit on stands, but most small snakes are handheld. These small-diameter snakes are not nearly as dangerous as their big brothers, but they do deserve your respect. These units carry all the normal hazards of an electric tool, but there is more.

The cables used for small drain-cleaning jobs are usually

very flexible. They are basically springs. The heads attached to the cables take on different shapes and looks. When you look at these thin cables, they don't look dangerous, but they can be. When the cables are being fed down a drain and turning, they can hit hard stoppages and go out of control. The cable can twist and kink. If your finger, hand, or arm is caught in the cable, injury can result. To avoid this, don't allow excessive cable between the drainpipe and the machine.

Large sewer machines

Large sewer machines are much more dangerous than small ones. These machines have tremendous power, and their cables are capable of removing fingers. Broken bones, severe cuts, and assorted other injuries are also possible with big snakes.

One of the most common problems with large sewer machines lies in the cables. When the cutting heads hit roots or similar hard items in the pipe, the cable goes wild. The twisting, thrashing cable can do a lot of damage. Again, limiting excess cable is one of the best forms of protection possible. Special sleeves are also available to contain unruly cables.

Most big machines can be operated by one person, but it is wise to have someone standing by in case help is needed. Loose clothing causes many drain-cleaning machine accidents. Wearing special mitts can help reduce the risk of hand injuries. Electric shocks are also possible when you do drain cleaning.

Power pipe threaders

Power pipe threaders are very good to have if you are doing much work with threaded pipe. However, these threading machines can grind body parts as well as they thread pipe. Electric threaders are very dangerous in the hands of untrained people. It is critical to keep fingers and clothing

away from the power mechanisms. The metal shavings produced by pipe threaders can be very sharp, and burrs left
on the threaded pipe can slash your skin. The cutting oil
used to keep the dies from getting too hot can make the
floor around the machine slippery.

Air-powered tools

Air-powered tools are not used often by plumbers.
Jackhammers are probably the most used air-powered tool
for plumbers. When you are using tools with air hoses,
check all connections carefully. If you experience a blowout,
the hose can spiral wildly out of control.

Powder-actuated tools

Powder-actuated tools are used by plumbers to secure
objects to hard surfaces, such as concrete. If the user is
properly trained, these tools are not too dangerous.
However, good training, eye protection, and ear protection
are all necessary. Misfires and chipping hard surfaces are
the most common danger with these tools.

Ladders

Ladders are used frequently by plumbers, both stepladders
and extension ladders (Fig. 15.6). Many ladder accidents
are waiting to happen. You must always be aware of what
is around you when you handle a ladder. If you brush
against a live electric wire with a ladder you are carrying,
your life could be over. Ladders often fall over when the
people using them are not careful. Reaching too far from a
ladder can be all that it takes for you to fall.

Be sure to set up a ladder or rolling scaffolds (Fig. 15.7)
properly. The ladder should be on firm footing, and all safety braces and clamps must be in place. When using an
extension ladder, many plumbers use a rope to tie rungs
together where the sections overlap. The rope provides an

1. Set the ladder on a solid and level footing.
2. Use a ladder in good condition; do not use one that needs repair.
3. Be sure step ladders are opened fully and locked.
4. When you use an extension ladder, place it at least one-fourth of its length away from the base of the building.
5. Tie an extension ladder to the building or other support, to prevent it from falling or blowing over in high winds.
6. Extend a ladder at least 3 ft over the roof line.
7. Have both hands free when you climb a ladder.
8. Do not carry tools in your pocket when you climb a ladder. (If you fall, the tools could cut you and cause serious injury.)
9. Use the ladder as it should be used. For example, do not allow two people on a ladder designed for use by one person.
10. Keep the ladder and all its steps clean—free of grease, oil, mud, etc.—in order to avoid a fall and possible injury.

Figure 15.6   Working safely on a ladder.

1. Do not lay tools or other materials on the floor of a scaffold. They can easily move, and you could trip over them; or they might fall, hitting someone on the ground.
2. Do not move a scaffold while you are on it.
3. Always lock the wheels when the scaffold is positioned and you are using it.
4. Always keep the scaffold level to maintain a steady platform on which to work.
5. Take no shortcuts. Be watchful at all times, and be prepared for any emergencies.

Figure 15.7   Safety on rolling scaffolds.

extra guard against failure of the ladder's safety clamps and collapse of the ladder. When you use an extension ladder, be sure to secure both the base and the top. I had an unusual accident on a ladder that I would like to share with you.

I was on a tall extension ladder, working near the top of a commercial building. The top of the ladder was resting on the edge of the flat roof. There was metal flashing surrounding the edge of the roof, and the top of the ladder was leaning against the flashing. There was a picket fence behind me, and electric wires entered the building to my right. The entrance wires were a good distance away, so I was in no immediate danger. As I worked on the ladder, a huge gust of wind blew around the building. I don't know where it came from; it hadn't been very windy when I went up the ladder.

The wind hit me and pushed me and the ladder sideways. The top of the ladder slid easily along the metal flashing, and I couldn't grab anything to stop me. I knew the ladder was going to go down, and I didn't have much time to make a decision. If I pushed off the ladder, I would probably be impaled on the fence. If I rode the ladder down, it might hit the electric wires and fry me. I waited until the last minute and jumped off the ladder.

I landed on the wet ground with a thud, but I missed the fence. The ladder hit the wires and sparks flew. Fortunately, I wasn't hurt, and electricians were available to take care of the electrical problem. In this case, I wasn't really negligent, but I could have been killed. If I had secured the top of the ladder, the accident wouldn't have happened.

Screwdrivers and chisels

Eye injuries and puncture wounds are common when you work with screwdrivers and chisels. When the tools are used properly and safety glasses are worn, there are few accidents.

The key to avoiding injury with most hand tools is simply to use the right tool for the job. If you use a wrench as a hammer or a screwdriver as a chisel, you are asking for trouble.

There are, of course, other types of tools (for example, see Fig. 15.8) and safety hazards in the plumbing trade. However, these guidelines cover the hazards that result in the greatest number of injuries. In all cases, observe proper

1. Read the operating instructions before you use the grinder.
2. Do not wear any loose clothing or jewelry.
3. Wear safety glasses or goggles.
4. Do not wear gloves while using the machine.
5. Shut the machine off promptly when you have finished with it.

Figure 15.8 Safe operation of grinders.

safety procedures and utilize safety gear, such as eye and ear protection.

Coworker Safety

Coworker safety is the last segment of this chapter. I include it because workers are frequently injured by the actions of coworkers. This section is meant to protect you from others and to make you aware of how your actions might affect your coworkers.

Most plumbers work around other people. This is especially true on construction jobs. When working around other people, you must be aware of their actions as well as your own. If you are walking out of a house to get something from the truck and a roll of roofing paper gets away from a roofer, you could get an instant headache.

If you don't pay attention to what is going on around you, you may wind up in all sorts of trouble. Cranes lose their loads sometimes, and such a load landing on you is likely to be fatal. Equipment operators don't always see the plumber kneeling down for a piece of pipe. It's not hard to have a close encounter with heavy equipment. While we are on the subject of equipment, let me bore you with another war story.

One day I was in a sewer ditch, connecting the sewer from a new house to the sewer main. The section of ditch where I was working was only about 4 ft deep. There was a large pile of dirt near the edge of the trench; it had been created when the ditch was dug. The dirt wasn't laid back as it should have been; it was piled up.

As I worked in the ditch, a backhoe came by. The operator had no idea I was in the ditch. When he swung the backhoe around to make a turn, the small scorpion-type bucket on the back of the equipment hit the dirt pile.

I had stood up when I heard the backhoe approaching, and it was a good thing. When the equipment hit the pile of dirt, part of the mound caved in on me. I tried to run, but it caught both my legs, and the weight drove me to the ground. I was buried from just below my waist. My head was okay, and my arms were free. I was still holding my shovel.

I yelled, but nobody heard me. I must admit, I was a little panicked. I tried to get up and couldn't. After a while, I was able to move enough dirt with the shovel to crawl out from under the dirt. I was lucky. If I had been on my knees making the connection or grading the pipe, I might have been buried. As it was, I came out of the ditch no worse for the wear. But was I ever mad at the careless backhoe operator!

That accident is a prime example of how other workers can hurt you and never know it. You have to watch out for yourself at all times (Fig. 15.9). As you gain field experi-

---

1. Be careful of underground utilities when you are digging.
2. Do not allow people to stand on the top edge of a ditch while workers are in the ditch.
3. Shore all trenches deeper than 4 ft.
4. When you are digging a trench, be sure to throw the dirt away from the ditch walls.
5. Be careful that no water gets into a trench. Be especially careful in areas with a high water table. Water in a trench can easily undermine the trench walls and lead to a cave-in.
6. Never work alone in a trench.
7. Always have someone nearby, someone who can help you and get additional help.
8. Always keep a ladder nearby so you can exit the trench quickly, if need be.
9. Be watchful at all times. Be aware of any potentially dangerous situations. Remember, even heavy truck traffic nearby can cause a cave-in.

Figure 15.9 Rules for working safely in ditches or trenches.

ence, you will develop a second nature for impending coworker problems. You will learn to sense when something is wrong or is about to go wrong. But you have to stay alive and healthy long enough to get that experience.

Always be aware of what is going on over your head. Avoid working under other people and hazardous overhead conditions. Let people know where you are, so you won't get stranded on a roof or in an attic when your ladder is moved or falls over.

Also remember that your actions could harm coworkers. If you are on a roof to flash a pipe and the hammer gets away from you, somebody could get hurt. Open communication between workers is one of the best ways to avoid injuries. If everyone knows where everyone else is working, injuries are less likely. Primarily, think—then think some more. There is no substitute for common sense. Avoid working alone, and remain alert at all times.

16

First Aid Basics

Everyone should invest some time to learn the basics of first aid. You never know when having skills in first aid treatments may save your life. Plumbers live what can be dangerous lives. On-the-job injuries are common. Most injuries are fairly minor, but they often require treatment. Do you know the right way to get a sliver of copper out of your hand? If your helper gets an electric shock when a drill cord goes bad, do you know what to do? Well, many plumbers don't have good first aid skills.

Before we get too far into this chapter, there are a few points I want to make. First, I'm not a medical doctor or any type of trained medical-care person. I've taken first aid classes, but I'm certainly not an authority on medical issues. The suggestions that I offer in this chapter are for informational purposes only. This book is not a substitute for first aid training offered by qualified professionals.

My intent here is to make you aware of some basic first aid procedures that can make life on the job much easier. But do not use my advice as instructions in first aid. This chapter will show you the benefits of taking first aid classes. Before you attempt first aid on anyone, including yourself, attend a structured, approved first aid class. I'm going to give you information that is as accurate as I can make it, but don't assume that my words are enough. Take a little time to seek professional training in the art of first aid. You may never use what you learn, but the one time it is needed, you will be glad you knew what to do.

Open Wounds

Open wounds are a common problem for plumbers. Many tools and materials used by plumbers can create open wounds. What should you do if you or one of your workers is cut?

- Stop the bleeding as soon as possible.
- Disinfect and protect the wound from contamination.
- Take steps to avoid symptoms of shock.
- Once the patient is stable, seek medical attention for severe cuts.

When the cut is bad, the victim may slip into shock. A loss of consciousness could result from a loss of blood. Death from extreme bleeding is also a risk. As a first aid provider, you must act quickly to reduce the risk of serious complications.

Bleeding

To stop bleeding, direct pressure is normally a good tactic. This may be as crude as clamping your hand over the wound, but a cleaner compression is desirable. Ideally, a sterile material should be placed over the wound and secured, normally with tape (even if it's duct tape). Thick gauze used as a pressure material can absorb blood and allow the clotting process to start.

Bad wounds may bleed right through the compress material. If this happens, don't remove the blood-soaked material. Add a new layer of material over it. Keep pressure on the wound. If you are not prepared with a first aid kit, substitute gauze and tape with strips cut from clothing that can be tied in place over the wound.

Elevate it

When you are dealing with a bleeding wound, it is usually best to elevate it. If you suspect a fractured or broken bone in the area of the wound, elevation may not be practical. When we talk about elevating a wound, it simply means to raise the wound above the level of the victim's heart. This slows down the blood flow, because of gravity.

Superserious bleeding

Superserious bleeding might not stop even after a compression bandage is applied and the wound is elevated. When this is the case, put pressure on the main artery that is producing the blood. Constricting an artery is not an alternative for the steps that we have discussed previously.

Putting pressure on an artery is serious business. First, you must be able to locate the artery, and then not keep the artery constricted any longer than necessary. You may have to apply pressure for awhile, release it, and then apply it again. It's important not to restrict the flow of blood in arteries for long periods. I hesitate to go into too much detail about this process, as I feel it is a method that should be taught in a controlled, classroom situation. However, I will hit the high spots. But remember, these words are not a substitute for professional training from qualified instructors.

Open arm wounds are controlled with the brachial artery. This artery is located between the biceps and triceps, on the inside of the arm. It's about halfway between the armpit and the elbow. Create pressure with the flat parts of your fingertips. Basically, hold the victim's wrist with one hand and close off the artery with your other hand. Pressure exerted by your fingers pushes the artery against the arm bone and restricts blood flow. Again, don't attempt this type of first aid until you have been trained properly.

Severe leg wounds may require the constriction of the femoral artery. This artery is located in the pelvic region. Normally, bleeding victims are placed on their backs for this procedure. The heel of a hand is placed on the artery to restrict blood flow. In some cases, fingertips are used to apply pressure.

Tourniquets

Tourniquets get a lot of attention in movies, but they can do as much harm as good if they are not used properly. A tourniquet should be used only in a life-threatening situation. When a tourniquet is applied, one risks losing the limb to which the restriction is applied. This is obviously a serious decision and one that must be made only when all other means of stopping blood loss have been exhausted.

Unfortunately, plumbers might find themselves in a situation where a tourniquet is the only answer. For example, if a worker allowed a power saw to get out of control, a hand might be severed or some other type of life-threatening injury could occur. This would be a reason to use a tourniquet. Let me give a basic overview of what's involved when a tourniquet is used.

Tourniquets should be at least 2 in wide. A tourniquet should be placed at a point above the wound, between the site of the bleeding and the victim's heart. However, the binding should not encroach directly on the wound area. Tourniquets can be fashioned out of many materials. If you are using strips of cloth, wrap the cloth around the limb and tie a knot in the material. Use a stick, screwdriver, or whatever else you can lay your hands on to tighten the binding.

Once you have applied a tourniquet, the wrapping should be removed only by a physician. It's a good idea to note the time at which the tourniquet is applied, as this will help doctors later in assessing their options. As an extension of the tourniquet treatment, you will most likely have to treat the patient for shock.

Infection

Infection is always a concern with open wounds. When a wound is serious enough to require a compression bandage, don't try to clean the cut. Keep pressure on the wound to stop the bleeding. In cases of severe wounds, watch for symptoms of shock, and be prepared to treat them. Your primary concern with a serious open wound is to stop the bleeding and get professional medical help as soon as possible.

Lesser cuts, which are more common than deep ones, should be cleaned. Use soap and water to clean a wound before you apply a bandage. Remember, we are talking about minor cuts and scrapes at this point. Flush the wound generously with clean water. Use a piece of sterile gauze to pat the wound dry. Then apply a clean, dry bandage to protect the wound during transport to a medical facility.

Splinters and Such

Splinters and foreign objects often invade the skin of plumbers. Getting these items out cleanly is best done by a doctor, but there are some on-the-job methods that you might want to try. A magnifying glass and a pair of tweezers work well together to remove embedded objects, such as splinters and slivers of copper tubing. Ideally, tweezers should be sterilized either over an open flame, such as the torch, or in boiling water.

Splinters and slivers under the skin can often be lifted out with the tip of a sterilized needle. Use of a needle and a pair of tweezers is very effective in the removal of most simple splinters. If you are dealing with something that has gone extremely deep into tissue, it is best to leave it alone until a doctor can remove it.

Eye Injuries

Eye injuries are very common on construction and remodeling jobs. Most of these injuries could be avoided if proper eye protection were worn, but far too many workers fail to wear safety glasses and goggles. This sets the stage for eye irritations and injuries.

Before you try to help someone suffering from an eye injury, wash your hands thoroughly. I know this is not always possible on construction sites, but cleaning your hands is important. In the meantime, keep the victim from rubbing the injured eye. Rubbing can make matters much worse.

Never try to remove a foreign object from someone's eye by using a rigid device, such as a toothpick. Cotton swabs that have been wetted can serve well as a magnet to remove some objects. If the victim has something embedded in an eye, get the person to a doctor as soon as possible. Don't attempt to remove the object yourself.

When you are investigating the cause of an eye injury, pull down the lower lid of the eye to check for the object causing trouble. A floating object, such as a piece of sawdust trapped between the eye and eyelid, can be removed with a tissue, a damp cotton swab, or even a clean handkerchief. Don't allow dry cotton material to come into contact with an eye.

If looking under the lower lid doesn't reveal the source of the discomfort, check under the upper lid. Clean water can be used to flush out many eye contaminants without much risk of damage to the eye. Objects that cannot be removed easily should be left alone until a physician can take over.

Scalp Injuries

Scalp injuries can be misleading. What looks like a serious wound can be a fairly minor cut. On the other hand, what appears to be only a cut can involve a fractured skull. If you or someone around you sustains a scalp injury, such as having a hammer fall on your head from overhead, take it seriously. Don't attempt to clean the wound. Expect profuse bleeding.

If you don't suspect a skull fracture, raise the victim's head and shoulders to reduce bleeding. Try not to bend the neck. Put a sterile bandage over the wound, but don't apply excessive pressure. If there is a bone fracture, pressure could worsen the situation. Secure the bandage with gauze or some other material that you can wrap around it. Seek medical attention immediately.

Facial Injuries

Facial injuries can occur on plumbing jobs. I've seen helpers let their right-angle drills get away from them, with the result being hard knocks to the face. On one occasion, I remember a tooth being lost, and split lips and bitten tongues are common when a drill goes on a rampage.

Extremely bad facial injuries can cause a blockage of the victim's air passages. This, of course, is a very serious condition. It's critical that air passages be open at all times. If the person has broken teeth or dentures, remove them. Be careful not to jar the individual's spine if you have reason to believe there may be an injury to the back or neck.

Conscious victims should be positioned, when possible, so that secretions from the mouth and nose drain out. Shock is a potential concern in severe facial injuries. For most on-the-job injuries, plumbers should be made comfortable and sent to seek medical attention.

Nosebleeds

Nosebleeds are not usually difficult to treat. Typically, apply pressure to the side of the nose where there is bleeding. Applying cold compresses can also help. If the application of external pressure does not stop the bleeding, use a small, clean pad of gauze to create a dam on the inside of the nose. Then apply pressure on the outside of the nose. This will almost always work. If it doesn't, go to a doctor.

Back Injuries

There is really only one thing that you need to know about back injuries: *Don't move the injured party.* Call for professional help, and see that the victim remains still until help arrives. Moving someone who has suffered a back injury can be very risky. Don't move the victim unless it would be worse not to, such as when someone is trapped in a fire or some other deadly situation.

Leg and Foot Injuries

Legs and feet are sometimes injured on job sites. In the worst case that I witnessed, a plumber knocked over a pot of molten lead onto his foot. When someone suffers a minor foot or leg injury, clean and cover the wound. Bandages should be supportive without being constrictive. Elevate the appendage above the victim's heart when possible. Do not let the person walk. Remove boots and socks so that you can keep an eye on the victim's toes. If the toes begin to swell or turn blue, loosen the supportive bandages.

Blisters

Blisters may not seem like much of an emergency, but they can sure take the steam out of a helper or plumber. In most cases, blisters can be covered with a heavy gauze pad to reduce pain. It is generally recommended to leave blisters unbroken. When a blister breaks, the area should be cleaned and treated as an open wound. Some blisters tend to be more serious than others. For example, blisters in the palm of a hand or on the sole of a foot should be looked at by a doctor.

Hand Injuries

Hand injuries are common in the plumbing trade. Little cuts are the most frequent complaint. Getting flux in a cut is an eye-opening experience, so even the smallest break in the skin should be covered. Elevate serious hand injuries. This tends to reduce swelling. Do not try to clean really bad hand injuries. Use a pressure bandage to control bleeding. If the cut is on the palm of a hand, the victim can squeeze a roll of gauze to slow the flow of blood. Pressure should stop the bleeding, but if it doesn't, seek medical assistance. As with all injuries, use common sense on whether or not professional attention is needed after first aid is applied.

Shock

Shock is a condition that can be life-threatening even when the responsible injury is not otherwise fatal. We are talking about *traumatic shock*, not electric shock. Many factors can cause a person to go into shock. A serious injury is one common cause, but many others exist. There are certain signs of shock to look for.

If a person's skin turns pale or blue and is cold to the touch, it's a likely sign of shock. Skin that becomes moist and clammy can indicate shock. A general weakness is also a sign of shock. The pulse of someone going into shock is likely to exceed 100 beats per minute. Breathing is usually increased, but it may be shallow, deep, or irregular. Chest injuries usually result in shallow breathing. Victims who have lost blood may be thrashing about as they go into shock. Vomiting and nausea can also signal shock.

As a person slips into deeper shock, the individual may become unresponsive. Check the eyes; they may be widely dilated. Blood pressure can drop, and in time the victim will lose consciousness. Body temperature will fall, and death will be likely if treatment is not given.

There are three main goals when you treat someone for shock. (1) Get the person's blood circulating well. (2) Make sure the victim has an adequate supply of oxygen. (3) Maintain the person's body temperature.

When treating a person for shock, you should make the victim lie down. Cover the individual to minimize the loss of body heat. Get medical help as soon as possible. Keeping a person lying down helps the individual's blood circulate better. Remember, if you suspect back or neck injuries, don't move the person.

People who are unconscious should be placed on one side so that fluids will run out of the mouth and nose. It's also important to make sure that air passages are open. A person with a head injury may be laid out flat or propped up, but the head should not be lower than the rest of the body. It is sometimes advantageous to elevate the feet of a person in shock. However, if there is any difficulty of breathing or if pain increases when the feet are raised, lower them.

Body temperature is a big concern with people in shock. You want to overcome or avoid chilling. However, don't try to add heat to the surface of the person's body by artificial means. This can be damaging. Use only blankets, clothes, and other similar items to regain and maintain body temperature.

Avoid the temptation to offer the victim fluids, unless medical care is not going to be available for a long time. Avoid fluids completely if the person is unconscious or is subject to vomiting. Under most job site conditions, fluids should not be administered.

Burns

Burns are not common among plumbers, but they can occur in the workplace. There are three types of burns. First-degree burns are the least serious. These burns typically result from overexposure to the sun, which construction workers often suffer from; quick contact with a hot object, such as the tip of a torch; and scalding water, as in working with a boiler or water heater.

Second-degree burns are more serious. They can result

from a deep sunburn or from contact with hot liquids and flames. A person with a second-degree burn may have a red or mottled appearance, blisters, and wet-looking skin within the burn area. This wet look is due to a loss of plasma through the damaged layers of skin.

Third-degree burns are the most serious. They can be caused by contact with open flames, hot objects, or immersion in very hot water. Electrical injuries can also result in third-degree burns. This type of burn can look similar to a second-degree burn, but the difference is the loss of all layers of skin for third-degree burns.

Treatment

Treatment for most job-related burns can be administered on the job site and will not require hospitalization. First-degree burns should be washed with or submerged in cold water. A dry dressing can be applied if necessary. These burns are not too serious. Eliminating pain is the primary goal with first-degree burns.

Second-degree burns should be immersed in cold (but not icy) water. The soaking should continue for at least 1 hour and up to 2 hours. After soaking, layer the wound with clean cloths that have been dipped in ice water and wrung out. Then dry the wound by blotting, not rubbing. Apply, a dry, sterile gauze pad. Don't break open any blisters. Do not use ointments and sprays on severe burns. Burned arms and legs should be elevated, and medical attention should be sought.

Bad burns, the third-degree type, need quick medical attention. First, don't remove a burn victim's clothing—skin might come off with it. A thick, sterile dressing can be applied to the burn area. Personally, I would avoid this if possible. A dressing might stick to the mutilated skin and cause additional skin loss when the dressing is removed. When hands are burned, keep them elevated above the victim's heart. The same holds for feet and legs. Do not soak a third-degree burn in cold water; this could induce symptoms of shock. Don't use ointments, sprays, or other types of treatments. Get the burn victim to competent medical care as soon as possible.

Heat-Related Problems

Heat-related problems include heat stroke and heat exhaustion. Cramps are also possible when one is working in hot weather. Some people don't consider heat stroke to be serious, but they are wrong. Heat stroke can be life-threatening. People affected by heat stroke can develop body temperatures in excess of 106°F. Their skin is likely to be hot, red, and dry. You might think they would sweat, but they don't. The pulse is rapid and strong, and victims can sink into an unconscious state.

If you are dealing with heat stroke, you need to lower the person's body temperature quickly. There is a risk, however, of cooling the body too quickly once the victim's temperature is below 102°F. You can lower body temperature by using rubbing alcohol, cold packs, cold water on clothes, or immersion in a bathtub of cold water. Avoid using ice in the cooling process. Fans and air-conditioned space can be used to achieve your cooling goals. Reduce the body temperature to at least 102°F, and then go for medical help.

Cramps

Cramps are not uncommon among workers during hot spells. A simple massage can be all it takes to cure this problem. Saltwater solutions are another way to control cramps. Mix 1 teaspoon of salt per glass of water, and have the victim drink half a glass about every 15 minutes.

Exhaustion

Heat exhaustion is more common than heat stroke. A person affected by heat exhaustion is likely to maintain a fairly normal body temperature. But the person's skin may be pale and clammy. Sweating may be very noticeable, and the individual will probably complain of feeling tired and weak. Headaches, cramps, and nausea may accompany the symptoms. In some cases, fainting may occur.

The saltwater treatment described for cramps will normally work with heat exhaustion. The victim should lie down and elevate the feet about 1 ft off the floor or bed. Loosen clothing, and use cool, wet cloths to increase comfort. If vomiting occurs, get the person to a hospital for intravenous fluids.

We could continue talking about first aid for a much longer time. However, the help I can give you here for medical procedures is limited. You owe it to yourself, your family, and your coworkers to learn first aid techniques. This can be done best by attending formal classes in your area. Most towns and cities offer first aid classes on a regular basis. I strongly suggest that you enroll in one. Until you have some hands-on experience in a classroom and gain the depth of knowledge needed, you are not prepared for emergencies. Don't get caught short. Prepare now for the emergency that might never happen.

17

Conversion Charts and Tables

Plumbing work can involve some complicated math. While all plumbers should learn to do math on their own, there is no need to waste time making calculations in the field if you can refer to a table for the same information. This chapter is full of useful tables and formulas that will reduce the time spent doing mathematics in the field. The illustrations are easy to understand, and they are always ready to give you that quick help you need when dealing with common plumbing conversions. See Figs. 17.1 through 17.34.

Figure 17.1 Abbreviations

A or a	Area
AWG	American wire gauge
bbl	Barrels
B or b	Breadth
bhp	Brake horsepower
B.M.	Board measure
Btu	British thermal unit
BWG	Birmingham wire gauge
B & S	Brown and Sharpe wire gauge (American wire gauge)
C of g	Center of gravity
cond	Condensing
cu	Cubic
cyl	Cylinder
D or d	Depth or diameter
evap	Evaporation
F, F	Coefficient of friction; Fahrenheit
F or f	Force or factor of safety
ft · lb	Foot-pounds
gal	Gallons
H or h	Height or head of water
h, hr	Hour
hp	Horsepower
ihp	Indicated horsepower
L or l	Length
lb	Pound
lb/in^2, psi	Pounds per square inch
OD	Outside diameter (pipes)
oz	Ounce
pt	Pint
P or p	Pressure or load
R or r	Radius
rpm, r/min	Revolutions per minute
sq ft, ft^2	Square feet
sq in, in^2	Square inches
sq yd, yd^2	Square yards
T or t	Thickness or temperature
temp	Temperature
V or v	Velocity
vol	Volume
W or w	Weight
WI	Wrought iron

Figure 17.2 Expansion in Plastic Piping

The formula for calculating expansion or contraction in plastic piping is

$$L = Y\left(\frac{T - E}{10}\right)\left(\frac{L}{100}\right)$$

where E = expansion, in

Y = constant factor expressing inches of expansion per 100°F temperature change per 100 ft of pipe

T = maximum temperature, °F

F = minimum temperature, °F

L = length of pipe run, ft

Figure 17.3 Temperature Conversion

Temperature may be expressed in the Fahrenheit scale or the Celsius scale. To convert from Celsius to Fahrenheit or Fahrenheit to Celsius, use the following formulas:

$$°F = 1.8 \times °C + 32$$
$$°C = 0.55555555 \, (°F - 32)$$

Figure 17.4 Flow Rate Equivalents

$$1 \text{ gal/min (gpm)} = 0.134 \text{ ft}^3/\text{min}$$
$$1 \text{ ft}^3/\text{min (cfm)} = 448.8 \text{ gal/h (gph)}$$

Figure 17.5 Design Temperature

Outside design temperature = average of lowest recorded temperature in each month from October to March

Inside design temperature = 70°F or as specified by owner

A degree-day is 1 day × the number of Fahrenheit degrees that the mean temperature is below 65°F. The number of degree-days in 1 year is a good guideline for designing heating and insulation systems.

Figure 17.6   Converting from Customary (U.S.) Units to Metric Units

To find:	Multiply:	By:
Micrometers, microns	Mils	25.4
Centimeters	Inches	2.54
Meters	Feet	0.3048
Meters	Yards	0.19144
Kilometers	Miles	1.609344
Grams	Ounces	28.349523
Kilograms	Pounds	0.4539237
Liters	Gallons (U.S.)	3.7854118
Liters	Gallons (imperial)	4.546090
Milliliters (cubic centimeters)	Fluid ounces	29.573530
Milliliters (cubic centimeters)	Cubic inches	16.387064
Square centimeters	Square inches	6.4516
Square meters	Square feet	0.09290304
Square meters	Square yards	0.83612736
Cubic meters	Cubic feet	2.8316847×10^{-2}
Cubic meters	Cubic yards	0.76455486
Joules	Btu	1054.3504
Joules	Foot-pounds	1.35582
Kilowatts	Btu per minute	0.01757251
Kilowatts	Foot-pounds per minute	2.2597×10^{-5}
Kilowatts	Horsepower	0.7457
Radians	Degrees	0.017453293
Watts	Btu per minute	17.5725

Figure 17.7   Liquid Measurements

4 gills = 1 pint	= 28.875 cubic inches (in^3)
2 pints (pt) = 1 quart	= 57.75 cubic inches
4 quarts (qt) = 1 gallon (gal)	= 231 cubic inches

Figure 17.8 Length Conversions from Inches to Meters and Millimeters

Inches (in)	Meters (m)	Millimeters (mm)
$\frac{1}{8}$	0.003	3.17
$\frac{1}{4}$	0.006	6.35
$\frac{3}{8}$	0.010	9.52
$\frac{1}{2}$	0.013	12.6
$\frac{5}{8}$	0.016	15.87
$\frac{3}{4}$	0.019	19.05
$\frac{7}{8}$	0.022	22.22
1	0.025	25.39
2	0.051	50.79
3	0.076	76.20
4	0.102	101.6
5	0.127	126.9
6	0.152	152.4
7	0.178	177.8
8	0.203	203.1
9	0.229	228.6
10	0.254	253.9
11	0.279	279.3
12	0.305	304.8

Figure 17.9   Useful Formulas

1. Circumference of circle = π × diameter = 3.1416 × diameter
2. Diameter of circle = circumference × 0.31831
3. Area of square = length × width
4. Area of rectangle = length × width
5. Area of parallelogram = base × perpendicular height
6. Area of triangle = $\frac{1}{2}$ × base × perpendicular height
7. Area of circle = π × radius squared = diameter squared × 0.7854
8. Area of ellipse = length × width × 0.7854
9. Volume of cube or rectangular prism = length × width × height
10. Volume of triangular prism = area of triangle × length
11. Volume of sphere = diameter cubed × 0.5236 = dia. × dia. × dia. × 0.5236
12. Volume of cone = π × radius squared × $\frac{1}{3}$ × height
13. Volume of cylinder = π × radius squared × height
14. Length of side of square × 1.128 = diameter of an equal circle
15. Doubling the diameter of a pipe or cylinder increases its capacity 4 times
16. Pressure (in pounds per square inch) of column of water = height of column (in feet) × 0.434
17. Capacity of pipe or tank (in U.S. gallons) = diameter squared (in inches) × length (in inches) × 0.0034
18. 1 gal water = $8\frac{1}{3}$ lb = 231 in^3
19. 1 ft^3 water = $62\frac{1}{2}$ lb = $7\frac{1}{2}$ gal

Figure 17.10 Inches Converted to Decimals of Feet

Inches	Decimal of a foot
$\frac{1}{8}$	0.01042
$\frac{1}{4}$	0.02083
$\frac{3}{8}$	0.03125
$\frac{1}{2}$	0.04167
$\frac{5}{8}$	0.05208
$\frac{3}{4}$	0.06250
$\frac{7}{8}$	0.07291
1	0.08333
$1\frac{1}{8}$	0.09375
$1\frac{1}{4}$	0.10417
$1\frac{3}{8}$	0.11458
$1\frac{1}{2}$	0.12500
$1\frac{5}{8}$	0.13542
$1\frac{3}{4}$	0.14583
$1\frac{7}{8}$	0.15625
2	0.16666
$2\frac{1}{8}$	0.17708
$2\frac{1}{4}$	0.18750
$2\frac{3}{8}$	0.19792
$2\frac{1}{2}$	0.20833
$2\frac{5}{8}$	0.21875
$2\frac{3}{4}$	0.22917
$2\frac{7}{8}$	0.23959
3	0.25000

Note: To change inches to decimals of a foot, divide by 12. To change decimals of a foot to inches, multiply by 12.

Figure 17.11   Atmospheric Pressure

Barometer (in of mercury)	Pressure (lb/in^2, or psi)
28.00	13.75
28.25	13.88
28.50	14.00
28.75	14.12
29.00	14.24
29.25	14.37
29.50	14.49
29.75	14.61
29.921	14.696
30.00	14.74
30.25	14.86
30.50	14.98
30.75	15.10
31.00	15.23

Rule: Barometer in inches of mercury $\times$ 0.49116 = lb/in^2.

Figure 17.12 Decimal Equivalents of Fractions of an Inch

Inch	Decimal of an inch
1/64	0.015625
1/32	0.03125
3/64	0.046875
1/16	0.0625
5/64	0.078125
3/32	0.09375
7/64	0.109375
1/8	0.125
9/64	0.140625
5/32	0.15625
11/64	0.171875
3/16	0.1875
13/64	0.203125
7/32	0.21875
15/64	0.234375
1/4	0.25
17/64	0.265625
9/32	0.28125
19/64	0.296875
5/16	0.3125

Note: To find the decimal equivalent of a fraction, divide the numerator by the denominator.

Figure 17.13   Formulas for a Circle

Circumference = diameter × 3 .1416

Circumference = radius × 6.2832

Diameter = radius × 2

Diameter = square root of area ÷ 0.7854

Diameter = square root of area × 1.1283

Diameter = circumference × 0.31831

Radius = diameter ÷ 2

Radius = circumference × 0.15915

Radius = square root of area × 0.56419

Area = diameter × diameter × 0.7854

Area = half of circumference × half of diameter

Area = square of circumference × 0.0796

Arc length = degrees × radius × 0.01745

Degrees of arc = length ÷ (radius × 0.01745)

Radius of arc = length ÷ (degrees × 0.01745)

Side of equal square = diameter × 0.8862

Side of inscribed square = diameter × 0.7071

Area of sector = area of circle × degrees of arc ÷ 360

Figure 17.14   Measurement Conversion Factors

To change:	To:	Multiply by:
Inches	Feet	0.0833
Inches	Millimeters	25.4
Feet	Inches	12
Feet	Yards	0.3333
Yards	Feet	3
Square inches	Square feet	0.00694
Square feet	Square inches	144
Square feet	Square yards	0.11111
Square yards	Square feet	9
Cubic inches	Cubic feet	0.00058
Cubic feet	Cubic inches	1728
Cubic feet	Cubic yards	0.03703
Gallons	Cubic inches	231
Gallons	Cubic feet	0.1337
Gallons	Pounds of water	8.33
Pounds of water	Gallons	0.12004
Ounces	Pounds	0.0625
Pounds	Ounces	16
Inches of water	Pounds per square inch	0.0361
Inches of water	Inches of mercury	0.0735
Inches of water	Ounces per square inch	0.578
Inches of water	Pounds per square foot	5.2
Inches of mercury	Inches of water	13.6
Inches of mercury	Feet of water	1.1333
Inches of mercury	Pounds per square inch	0.4914
Ounces per square inch	Inches of mercury	0.127
Ounces per square inch	Inches of water	1.733
Pounds per square inch	Inches of water	27.72
Pounds per square inch	Feet of water	2.310
Pounds per square inch	Inches of mercury	2.04
Pounds per square inch	Atmospheres	0.0681
Feet of water	Pounds per square inch	0.434
Feet of water	Pounds per square foot	62.5
Feet of water	Inches of mercury	0.8824
Atmospheres	Pounds per square inch	14.696
Atmospheres	Inches of mercury	29.92
Atmospheres	Feet of water	34
Long tons	Pounds	2240
Short tons	Pounds	2000
Short tons	Long tons	0.89295

Figure 17.15   Conversion Table

0.001 inch (in) = 0.025 millimeter (mm)
1 in = 25.400 mm
1 foot (ft) = 30.48 centimeters (cm)
1 ft = 0.3048 meter (m)
1 yard (yd) = 0.9144 m
1 mile (mi) = 1.609 kilometers (km)
1 square in (in^2) = 6.4516 cm^2
1 ft^2 = 0.0929 m^2
1 yd^2 = 0.8361 m^2
1 acre = 0.4047 hectare (ha)
1 square mile (mi^2) = 2.590 km^2
1 cubic inch (in^3) = 16.387 cm^3
1 ft^3 = 0.0283 m^3
1 yd^3 = 0.7647 m^3
1 U.S. ounce (oz) = 29.57 milliliters (mL)
1 U.S. pint (pt) = 0.4732 liter (L)
1 U.S. gallon (gal) = 3.785 L
1 oz = 28.35 grams (g)
1 pound (lb) = 0.4536 kilogram (kg)

Figure 17.16 Circumference of a Circle

Diameter, in	Circumference, in
$\frac{1}{8}$	0.3927
$\frac{1}{4}$	0.7854
$\frac{3}{8}$	1.178
$\frac{1}{2}$	1.570
$\frac{5}{8}$	1.963
$\frac{3}{4}$	2.356
$\frac{7}{8}$	2.748
1	3.141
$1\frac{1}{8}$	3.534
$1\frac{1}{4}$	3.927
$1\frac{3}{8}$	4.319
$1\frac{1}{2}$	4.712
$1\frac{5}{8}$	5.105
$1\frac{3}{4}$	5.497
$1\frac{7}{8}$	5.890
2	6.283
$2\frac{1}{4}$	7.068
$2\frac{1}{2}$	7.854
$2\frac{3}{4}$	8.639
3	9.424
$3\frac{1}{4}$	10.21
$3\frac{1}{2}$	10.99
$3\frac{3}{4}$	11.78
4	12.56
$4\frac{1}{2}$	14.13
5	15.70
$5\frac{1}{2}$	17.27

Figure 17.17   Area of a Circle

Diameter, in	Area, in
$\frac{1}{8}$	0.0123
$\frac{1}{4}$	0.0491
$\frac{3}{8}$	0.1104
$\frac{1}{2}$	0.1963
$\frac{5}{8}$	0.3068
$\frac{3}{4}$	0.4418
$\frac{7}{8}$	0.6013
1	0.7854
$1\frac{1}{8}$	0.9940
$1\frac{1}{4}$	1.227
$1\frac{3}{8}$	1.484
$1\frac{1}{2}$	1.767
$1\frac{5}{8}$	2.073
$1\frac{3}{4}$	2.405
$1\frac{7}{8}$	2.761
2	3.141
$2\frac{1}{4}$	3.976
$2\frac{1}{2}$	4.908
$2\frac{3}{4}$	5.939
3	7.068
$3\frac{1}{4}$	8.295
$3\frac{1}{2}$	9.621
$3\frac{3}{4}$	11.044
4	12.566
$4\frac{1}{2}$	15.904
5	19.635
$5\frac{1}{2}$	23.758

Figure 17.18   Useful Multipliers

To change:	To:	Multiply by:
Inches	Feet	0.0833
Inches	Millimeters	25.4
Feet	Inches	12
Feet	Yards	0.3333
Yards	Feet	3
Square inches	Square feet	0.00694
Square feet	Square inches	144
Square feet	Square yards	0.11111
Square yards	Square feet	9
Cubic inches	Cubic feet	0.00058
Cubic feet	Cubic inches	1728
Cubic feet	Cubic yards	0.03703
Cubic yards	Cubic feet	27
Cubic inches	Gallons	0.00433
Cubic feet	Gallons	7.48
Gallons	Cubic inches	231
Gallons	Cubic feet	0.1337
Gallons	Pounds of water	8.33
Pounds of water	Gallons	0.12004
Ounces	Pounds	0.0625
Pounds	Ounces	16
Inches of water	Pounds per square inch	0.0361
Inches of water	Inches of mercury	0.0735
Inches of water	Ounces per square inch	0.578
Inches of water	Pounds per square foot	5.2
Inches of mercury	Inches of water	13.6
Inches of mercury	Feet of water	1.1333
Ounces per square inch	Pounds per square inch	0.127
Ounces per square inch	Inches of mercury	1.733
Pounds per square inch	Inches of water	27.72
Pounds per square inch	Feet of water	2.310
Pounds per square inch	Inches of mercury	2.04
Pounds per square inch	Atmospheres	0.0681
Feet of water	Pounds per square inch	0.434
Feet of water	Pounds per square foot	62.5
Feet of water	Inches of mercury	0.8824
Atmospheres	Pounds per square inch	14.696
Atmospheres	Inches of mercury	29.92
Atmospheres	Feet of water	34
Long tons	Pounds	2240
Short tons	Pounds	2000
Short tons	Long tons	0.89285

Figure 17.19 Square Roots of Numbers

Number	Square root	Number	Square root
1	1.00000	31	5.56776
2	1.41421	32	5.65685
3	1.73205	33	5.74456
4	2.00000	34	5.83095
5	2.23606	35	5.91607
6	2.44948	36	6.00000
7	2.64575	37	6.08276
8	2.82842	38	6.16441
9	3.00000	39	6.24499
10	3.16227	40	6.32455
11	3.31662	41	6.40312
12	3.46410	42	6.48074
13	3.60555	43	6.55743
14	3.74165	44	6.63324
15	3.87298	45	6.70820
16	4.00000	46	6.78233
17	4.12310	47	6.85565
18	4.24264	48	6.92820
19	4.35889	49	7.00000
20	4.47213	50	7.07106
21	4.58257	51	7.14142
22	4.69041	52	7.21110
23	4.79583	53	7.28010
24	4.89897	54	7.34846
25	5.00000	55	7.41619
26	5.09901	56	7.48331
27	5.19615	57	7.54983
28	5.29150	58	7.61577
29	5.38516	59	7.68114
30	5.47722	60	7.74596

Figure 17.19 Square Roots of Numbers (*Continued*)

Number	Square root	Number	Square root
61	7.81024	81	9.00000
62	7.87400	82	9.05538
63	7.93725	83	9.11043
64	8.00000	84	9.16515
65	8.06225	85	9.21954
66	8.12403	86	9.27361
67	8.18535	87	9.32737
68	8.24621	88	9.38083
69	8.30662	89	9.43398
70	8.36660	90	9.48683
71	8.42614	91	9.53939
72	8.48528	92	9.59166
73	8.54400	93	9.64365
74	8.60232	94	9.69535
75	8.66025	95	9.74679
76	8.71779	96	9.79795
77	8.77496	97	9.84885
78	8.83176	98	9.89949
79	8.88819	99	9.94987
80	8.94427	100	10.00000

Figure 17.20   Cubes of Numbers

Number	Cube	Number	Cube
1	1	31	29,791
2	8	32	32,768
3	27	33	35,937
4	64	34	39,304
5	125	35	42,875
6	216	36	46,656
7	343	37	50,653
8	512	38	54,872
9	729	39	59,319
10	1,000	40	64,000
11	1,331	41	68,921
12	1,728	42	74,088
13	2,197	43	79,507
14	2,477	44	85,184
15	3,375	45	91,125
16	4,096	46	97,336
17	4,913	47	103,823
18	5,832	48	110,592
19	6,859	49	117,649
20	8,000	50	125,000
21	9,621	51	132,651
22	10,648	52	140,608
23	12,167	53	148,877
24	13,824	54	157,464
25	15,625	55	166,375
26	17,576	56	175,616
27	19,683	57	185,193
28	21,952	58	195,112
29	24,389	59	205,379
30	27,000	60	216,000

Figure 17.20   Cubes of Numbers (*Continued*)

Number	Cube	Number	Cube
61	226,981	81	531,441
62	238,328	82	551,368
63	250,047	83	571,787
64	262,144	84	592,704
65	274,625	85	614,125
66	287,496	86	636,056
67	300,763	87	658,503
68	314,432	88	681,472
69	328,500	89	704,969
70	343,000	90	729,000
71	357,911	91	753,571
72	373,248	92	778,688
73	389,017	93	804,357
74	405,224	94	830,584
75	421,875	95	857,375
76	438,976	96	884,736
77	456,533	97	912,673
78	474,552	98	941,192
79	493,039	99	970,299
80	512,000	100	1,000,000

Figure 17.21   Squares of Numbers

Number	Square	Number	Square
1	1	31	961
2	4	32	1,024
3	9	33	1,089
4	16	34	1,156
5	25	35	1,225
6	36	36	1,296
7	49	37	1,369
8	64	38	1,444
9	81	39	1,521
10	100	40	1,600
11	121	41	1,681
12	144	42	1,764
13	169	43	1,849
14	196	44	1,936
15	225	45	2,025
16	256	46	2,116
17	289	47	2,209
18	324	48	2,304
19	361	49	2,401
20	400	50	2,500
21	441	51	2,601
22	484	52	2,704
23	529	53	2,809
24	576	54	2,916
25	625	55	3,025
26	676	56	3,136
27	729	57	3,249
28	784	58	3,364
29	841	59	3,481
30	900	60	3,600

Figure 17.21 Squares of Numbers (*Continued*)

Number	Square	Number	Square
61	3,721	81	6,561
62	3,844	82	6,724
63	3,969	83	6,889
64	4,096	84	7,056
65	4,225	85	7,225
66	4,356	86	7,396
67	4,489	87	7,569
68	4,624	88	7,744
69	4,761	89	7,921
70	4,900	90	8,100
71	5,041	91	8,281
72	5,184	92	8,464
73	5,329	93	8,649
74	5,476	94	8,836
75	5,625	95	9,025
76	5,776	96	9,216
77	5,929	97	9,409
78	6,084	98	9,604
79	6,241	99	9,801
80	6,400	100	10,000

Figure 17.22 Flow Rate Conversion from
Gallons per Minute (gpm) to Approximate
Liters per Minute (L/min)

gpm	L/min
1	3.75
2	6.50
3	11.25
4	15.00
5	18.75
6	22.50
7	26.25
8	30.00
9	33.75
10	37.50

Figure 17.23 Decimals to Millimeters

Decimal equivalent	Millimeters
0.0625	1.59
0.1250	3.18
0.1875	4.76
0.2500	6.35
0.3125	7.94
0.3750	9.52
0.4375	11.11
0.5000	12.70
0.5625	14.29
0.6250	15.87
0.6875	17.46
0.7500	19.05
0.8125	20.64
0.8750	22.22
0.9375	23.81
1.000	25.40

Figure 17.24 Decimal Equivalents of an Inch

$$\frac{1}{32} = 0.03125$$
$$\frac{1}{16} = 0.0625$$
$$\frac{3}{32} = 0.09375$$
$$\frac{1}{8} = 0.125$$
$$\frac{5}{32} = 0.15625$$
$$\frac{3}{16} = 0.1875$$
$$\frac{7}{32} = 0.21875$$
$$\frac{1}{4} = 0.25$$
$$\frac{9}{32} = 0.28125$$
$$\frac{5}{16} = 0.3125$$
$$\frac{11}{32} = 0.34375$$
$$\frac{3}{8} = 0.375$$
$$\frac{13}{32} = 0.40625$$
$$\frac{7}{16} = 0.4375$$
$$\frac{15}{32} = 0.46875$$
$$\frac{1}{2} = 0.5$$
$$\frac{17}{32} = 0.53125$$
$$\frac{9}{16} = 0.5625$$
$$\frac{19}{32} = 0.59375$$
$$\frac{5}{8} = 0.625$$
$$\frac{21}{32} = 0.65625$$
$$\frac{11}{16} = 0.6875$$
$$\frac{23}{32} = 0.71875$$
$$\frac{3}{4} = 0.75$$
$$\frac{25}{32} = 0.78125$$
$$\frac{13}{16} = 0.8125$$
$$\frac{27}{32} = 0.84375$$
$$\frac{7}{8} = 0.875$$
$$\frac{29}{32} = 0.90625$$
$$\frac{15}{16} = 0.9375$$
$$\frac{31}{32} = 0.96875$$
$$1 = 1.000$$

Figure 17.25 Fractions to Decimals

Fractions	Decimal equivalent
$1/16$	0.0625
$1/8$	0.1250
$3/16$	0.1875
$1/4$	0.2500
$5/16$	0.3125
$3/8$	0.3750
$7/16$	0.4375
$1/2$	0.5000
$9/16$	0.5625
$5/8$	0.6250
$11/16$	0.6875
$3/4$	0.7500
$13/16$	0.8125
$7/8$	0.8750
$15/16$	0.9375
1	1.000

Figure 17.26 Decimal Equivalents of Fractions

Fraction	Decimal
1/64	0.015625
1/32	0.03125
3/64	0.046875
1/20	0.05
1/16	0.0625
1/13	0.0769
5/64	0.078125
1/12	0.0833
1/11	0.0909
3/32	0.09375
1/10	0.10
7/64	0.109375
1/9	0.111
1/8	0.125
9/64	0.140625
1/7	0.1429
5/32	0.15625
1/6	0.1667
11/64	0.171875
3/16	0.1875
1/5	0.2
13/64	0.203125
7/32	0.21875
15/64	0.234375
1/4	0.25
17/64	0.265625
9/32	0.28125
19/64	0.296875

Figure 17.26   Decimal Equivalents of Fractions (*Continued*)

Fraction	Decimal
$5/16$	0.3125
$21/64$	0.328125
$1/3$	0.333
$11/32$	0.34375
$23/64$	0.359375
$3/8$	0.375
$25/64$	0.390625
$13/32$	0.40625
$27/64$	0.421875
$7/16$	0.4375
$29/64$	0.453125
$15/32$	0.46875
$31/64$	0.484375
$1/2$	0.5
$33/64$	0.515625
$17/32$	0.53125
$35/64$	0.546875
$9/16$	0.5625
$37/64$	0.578125
$19/32$	0.59375
$39/64$	0.609375
$5/8$	0.625
$41/64$	0.640625
$21/32$	0.65625
$43/64$	0.671875
$11/16$	0.6875
$45/64$	0.703125

Figure 17.27 Water Feet Head to Pounds per Square Inch

Feet head	Pounds per square inch
1	0.43
2	0.87
3	1.30
4	1.73
5	2.17
6	2.60
7	3.03
8	3.46
9	3.90
10	4.33
15	6.50
20	8.66
25	10.83
30	12.99
40	17.32
50	21.65
60	25.99
70	30.32
80	34.65
90	38.98
100	43.34
110	47.64
120	51.97
130	56.30
140	60.63
150	64.96
160	69.29
170	73.63
180	77.96
200	86.62

Figure 17.28   Water Pressure in Pounds with Equivalent Feet Head

Pounds per square inch	Feet head
1	2.31
2	4.62
3	6.93
4	9.24
5	11.54
6	13.85
7	16.16
8	18.47
9	20.78
10	23.09
15	34.63
20	46.18
25	57.72
30	69.27
40	92.36
50	115.45
60	138.54
70	161.63
80	184.72
90	207.81
100	230.90

Figure 17.29 Water Pressure to Feet Head

Water pressure	Feet head
90	207.81
100	230.90
110	253.98
120	277.07
130	300.16
140	323.25
150	346.34
160	369.43
170	392.52
180	415.61
200	461.78
250	577.24
300	692.69
350	808.13
400	922.58
500	1154.48
600	1385.39
700	1616.30
800	1847.20
900	2078.10
1000	2309.00

Figure 17.30 Common Square Measures

$$144 \text{ in}^2 = 1 \text{ ft}^2$$
$$9 \text{ ft}^2 = 1 \text{ yd}^2$$
$$1 \text{ yd}^2 = 1296 \text{ in}^2$$
$$4840 \text{ yd}^2 = 1 \text{ acre}$$
$$640 \text{ acres} = 1 \text{ mi}^2$$

Figure 17.31 Other Square Measures

$1\ cm^2 = 0.1550\ in^2$
$1\ dm^2 = 0.1076\ ft^2$
$1\ m^2 = 1.196\ yd^2$
$1\ are = 3.954\ rd^2$
$1\ hectare\ (ha) = 2.47\ acres$
$1\ km^2 = 0.386\ mi^2$
$1\ in^2 = 6.452\ cm^2$
$1\ ft^2 = 9.2903\ dm^2$
$1\ yd^2 = 0.8361\ m^2$
$1\ rd^2 = 0.2529\ are$
$1\ acre = 0.4047\ ha$
$1\ mi^2 = 2.59\ km^2$

Figure 17.32 Water Volume to Weight

$1\ ft^3 = 62.4\ lb$
$1\ ft^3 = 7.48\ gal$
$1\ gal = 8.33\ lb$
$1\ gal = 0.1337\ ft^3$

Figure 17.33   Measurement Conversions: Imperial to Metric

Length
1 in = 25.4 mm
1 ft = 0.3048 m
1 yd = 0.9144 m
1 mi = 1.609 km

Mass
1 lb = 0.454 kg
1 U.S. short ton = 0.9072 metric ton

Area
1 ft^2 = 0.092 m^2
1 yd^2 = 0.836 m^2
1 acre = 0.404 hectare (ha)

Capacity
1 ft^3 = 0.028 m^3

Or
1 yd^3 = 0.764 m^3

Volume
1 qt (liquid) = 0.946 liter (L)
1 gal = 3.785 L

Heat
1 Btu = 1055 joules (J)
1 Btu/h = 0.293 watt (W)

Figure 17.34 English-to-Metric Conversions

Inches $\times$ 25.4 = Millimeters
Feet $\times$ 0.3048 = Meters
Miles $\times$ 1.6093 = Kilometers
Square inches $\times$ 6.4515 = Square centimeters
Square feet $\times$ 0.09290 = Square meters
Acres $\times$ 0.4047 = Hectares
Acres $\times$ 0.00405 = Square kilometers
Cubic inches $\times$ 16.3872 = Cubic centimeters
Cubic feet $\times$ 0.02832 = Cubic meters
Cubic yards $\times$ 0.76452 = Cubic meters
Cubic inches $\times$ 0.01639 = Liters
U.S. gallons $\times$ 3.7854 = Liters
Ounces (avoirdupois) $\times$ 28.35 = Grams
Pounds $\times$ 0.4536 = Kilograms
Pounds per square inch (psi) $\times$ 0.0703 = Kilograms per square centimeter
Pounds per cubic foot $\times$ 16.0189 = Kilograms per cubic meter
Tons (2000 lb) $\times$ 0.0972 = Metric tons (1000 kg)
Horsepower $\times$ 0.746 = Kilowatts (kW)

18

Sensible Pipe Sizing

Sensible pipe sizing turns off many plumbers. Few plumbers enjoy sizing pipe systems. Part of the explanation may be the complicated methods used by some people to size systems. Does sizing have to be complicated? Is it really necessary to use friction-loss charts to size a water distribution system? What do you think?

I don't see any reason for pipe sizing to be difficult. Charts and tables in code books can be troublesome to understand. But sizing doesn't have to be a dreadful chore. I admit, at times friction charts and complicated formulas must be used to figure the proper sizing of a system. However, this is almost never the case in a residence, and commercial jobs are usually designed by engineers and architects, not plumbers. Let me show you some of the simple techniques that I use when sizing systems.

Pipe Sizing

Sizing pipe for a drainage system is not difficult. To size pipe for drainage, there are a few benchmark numbers you must know, but you don't have to memorize them. Your code book contains charts and tables that provide the benchmarks. All you must know is how to interpret and use the information found there.

Figure 18.1   Fixture-Unit Ratings in Zone 3

Bathtub	2
Shower	2
Residential toilet	4
Lavatory	1
Kitchen sink	2
Dishwasher	2
Clothes washer	3
Laundry tub	2

The size of a drainage pipe is determined by using various factors, the first of which is the *drainage load*. This refers to the volume of drainage that the pipe will be responsible for carrying. In the code book, you will find ratings that assign a fixture-unit value to various plumbing fixtures (Fig. 18.1). For example, a residential toilet has a fixture-unit value of 4. A bathtub's fixture-unit value is 2.

Using the ratings given in the code book, you can quickly assess the drainage load for the system you are designing. Since plumbing fixtures require traps, you must also determine what size traps are needed for particular fixtures. Again, you don't need a math degree to do this. In fact, the code book specifies the trap sizes required for most common plumbing fixtures.

Your code book lists trap size requirements for specific fixtures (Figs. 18.2, 18.3, and 18.4). For example, referring to the ratings in the code book, you will find that a bathtub requires a $1\frac{1}{2}$-in trap. A lavatory may have a $1\frac{1}{4}$-in trap. The list describes the trap needs for all common plumbing fixtures. Trap sizes are not given for toilets, since toilets have integral traps.

When necessary, you can determine a fixture's drainage unit value by the size of its trap. A $1\frac{1}{4}$-in trap, the smallest trap allowed, has a fixture-unit rating of 1. A $1\frac{1}{2}$-in trap has a fixture-unit rating of 2. A 2-in trap has a rating of 3 fixture units. A 3-in trap has a fixture-unit rating of 5, and a 4-in trap has a fixture-unit rating of 6. This information can be found in your code book (Figs. 18.5, 18.6, and 18.7)

Figure 18.2 Recommended Trap Sizes for Zone 1

Type of fixture	Trap size, in
Bathtub	$1\frac{1}{2}$
Shower	2
Residential toilet	Integral
Lavatory	$1\frac{1}{4}$
Bidet	$1\frac{1}{2}$
Laundry tub	$1\frac{1}{2}$
Washing machine standpipe	2
Floor drain	2
Kitchen sink	$1\frac{1}{2}$
Dishwasher	$1\frac{1}{2}$
Drinking fountain	$1\frac{1}{4}$
Public toilet	Integral

Figure 18.3 Recommended Trap Sizes for Zone 2

Type of fixture	Trap size, in
Bathtub	$1\frac{1}{2}$
Shower	2
Residential toilet	Integral
Lavatory	$1\frac{1}{4}$
Bidet	$1\frac{1}{2}$
Laundry tub	$1\frac{1}{2}$
Washing machine standpipe	2
Floor drain	2
Kitchen sink	$1\frac{1}{2}$
Dishwasher	$1\frac{1}{2}$
Drinking fountain	1
Public toilet	Integral

Figure 18.4 Recommended Trap Sizes for Zone 3

Type of fixture	Trap size, in
Bathtub	$1\frac{1}{2}$
Shower	2
Residential toilet	Integral
Lavatory	$1\frac{1}{4}$
Bidet	$1\frac{1}{4}$
Laundry tub	$1\frac{1}{2}$
Washing machine standpipe	2
Floor drain	2
Kitchen sink	$1\frac{1}{2}$
Dishwasher	$1\frac{1}{2}$
Drinking fountain	$1\frac{1}{4}$
Public toilet	Integral
Urinal	2

Figure 18.5 Fixture-Unit
Requirements on Trap Sizes in Zone 1

Trap size, in	Fixture units
$1\frac{1}{4}$	1
$1\frac{1}{2}$	3
2	4
3	6
4	8

Figure 18.6 Fixture-Unit
Requirements on Trap Sizes in Zone 2

Trap size, in	Fixture units
$1\frac{1}{4}$	1
$1\frac{1}{2}$	2
2	3
3	5
4	6

Figure 18.7 Fixture-Unit
Requirements on Trap Sizes

Trap size, in	Fixture units
$1\frac{1}{4}$	1
$1\frac{1}{2}$	2
2	3
3	5
4	6

and may be applied to a fixture not specifically given a rating in the code book. Remember, the three major plumbing codes are similar, but they are not identical. Also, plumbing codes change from time to time. Consult your local code for current, correct information.

Determining the fixture-unit value of a pump does require a little math, but it's simple. Assign 2 fixture units for every gallon per minute (gpm) of flow. For example, a pump with a flow rate of 30 gpm has a fixture-unit rating of 60. In zone 3, it is more generous. In zone 3, for every $7\frac{1}{2}$ gpm, 1 fixture unit is assigned. With the same pump, producing 30 gpm, in zone 3 the fixture-unit rating is 4. That's quite different from the ratings in zones 1 and 2.

To size drainpipe, you must also consider the type of drain and the amount of fall of the pipe. For example, the sizing for a sewer is done a little differently from the sizing for a vertical stack. A pipe with a $\frac{1}{4}$-in fall is rated differently from the same pipe with a $\frac{1}{8}$-in fall.

Sizing Building Drains and Sewers

In building drains and sewers (Fig. 18.8), the same factors determine the proper pipe size. To size these types of pipes, you must know the total number of drainage fixture units entering the pipe and the amount of fall placed on the pipe. The amount of fall is the distance the pipe drops in each foot of travel. A normal grade is generally $\frac{1}{4}$ in/ft, but the fall could be more or less.

Figure 18.8 Building Drain Size for Zone 3

Pipe size, in	Pipe grade, in/ft	Maximum number of fixture units
2	$\frac{1}{4}$	21
3	$\frac{1}{4}$	42*
4	$\frac{1}{4}$	216

Note: No more than two toilets may be installed on a 3-in building drain.

When you refer to the code book, you will find information, probably a table, to help you size building drains and sewers. Let's take a look at how a building drain for a typical house is sized in zone 3.

Sizing Example

Our sample house has $2\frac{1}{2}$ bathrooms, a kitchen, and a laundry room. To size the building drain for this house, we must determine the total fixture-unit load that could be placed on the building drain. To do this, we start by listing all the plumbing fixtures producing a drainage load. In this house we have the following fixtures:

One bathtub	One shower
Three toilets	Three lavatories
One kitchen sink	One dishwasher
One clothes washer	One laundry tub

Using a chart in the code book, we can determine the number of drainage fixture units assigned to each of these fixtures. When we add all the fixture units, the total load is 28 fixture units. It is always best to allow a little extra in the fixture-unit load so that the pipe will be in no danger of becoming overloaded. The next step is to look at a different chart in the code book to determine the size of the building drain.

The building drain will be installed with a $\frac{1}{4}$-in fall. Looking at the chart, we see that we can use a 3-in pipe for the building drain, based on the number of fixture units;

but there is a footnote below the chart. It indicates that a 3-in pipe may not carry the discharge of more than two toilets, and our test house has three toilets. So we have to move up to a 4-in pipe.

Suppose the test house had only two toilets. What then? If we eliminate one of the toilets, the fixture load drops to 24. According to the table, we could use a 2½-in pipe, but the building drain must be at least a 3-in pipe, to connect to the toilets. A fixture's drain may enter a pipe the same size as the fixture drain or a pipe that is larger, but it may never be reduced to a smaller size, except with a 4-in by 3-in closet bend.

So, with two toilets, our sample house could have a building drain and sewer with a 3-in diameter. But should we run a 3-in pipe or a 4-in pipe? In a highly competitive bidding situation, 3-in pipe would probably win the coin toss. It would be less expensive to install a 3-in drain, and we would be more likely to win the bid on the job. However, when feasible, it would be better to use a 4-in drain. This allows the homeowner to add another toilet sometime in the future. If we install a 3-in sewer, the homeowner cannot add a toilet without replacing the sewer with 4-in pipe.

Horizontal Branches

Horizontal branches are the pipes branching off from a stack, to accept the discharge from fixture drains. These normally leave the stack as horizontal pipe, but they may turn to a vertical position, while retaining the name *horizontal branch*.

The procedure for sizing a horizontal branch (Fig. 18.9) is similar to that used to size a building drain or sewer, but the ratings are different. The code book contains the benchmarks you need to size, but let me give you some examples.

The number of fixture units allowed on a horizontal branch is determined by the pipe size and the pitch. All the following examples are based on a pitch of ¼ in/ft. A 2-in pipe can accommodate up to 6 fixture units, except in zone

Figure 18.9   Example of Sizing Horizontal Branch in Zone 2

Pipe size, in	Maximum number of fixture units on horizontal branch
$1\frac{1}{4}$	1
$1\frac{1}{2}$	3
2	6
3	20*
4	160
6	620

Note: Not more than two toilets may be connected to a single 3-in horizontal branch. Any branch connecting with a toilet must have a minimum diameter of 3 in.

Note: Table does not represent branches of the building drain, and other restrictions apply under battery-venting conditions.

1, where it can have 8 fixture units. A 3-in pipe can handle 20 fixture units, but not more than two toilets. In zone 1, a 3-in pipe is allowed up to 35 fixture units and up to three toilets. A $1\frac{1}{2}$-in pipe carries 3 fixture units, unless you are in zone 1. In zone 1, only $1\frac{1}{2}$-in pipe carries 2 fixture units, and they may not be from sinks, dishwashers, or urinals. A 4-in pipe takes up to 160 fixture units, except in zone 1, where it takes up to 216 units.

Stack Sizing

Stack sizing is not too different from the other sizing exercises we have studied. When you size a stack (Figs. 18.10, 18.11, 18.12, and 18.13), you base your decision on the total number of fixture units carried by the stack and the amount of discharge into branch intervals. This may sound complicated, but it isn't.

When you look at a chart for stack sizing, note that there are three columns. The first is for the pipe size, the second represents the discharge of a branch interval, and the third shows the ratings for the total fixture-unit load on a stack. The table is based on a stack with no more than three branch intervals.

Figure 18.10   Stack Sizing for Zone 2

Pipe size, in	Fixture-unit discharge on stack from a branch	Total fixture units allowed on stack
$1\frac{1}{2}$	3	4
2	6	10
3	20*	30*
4	160	240

Note: No more than two toilets may be placed on a 3-in branch, and no more than six toilets may be connected to a 3-in stack.

Figure 18.11   Sizing of Tall Stacks in Zone 2

(Stacks with More than Three Branch Intervals)

Pipe size, in	Fixture-unit discharge on stack from a branch	Total fixture units allowed on stack
$1\frac{1}{2}$	2	8
2	6	24
3	16*	60*
4	90	500

Note: No more than two toilets may be placed on a 3-in branch, and no more than six toilets may be connected to a 3-in stack.

Figure 18.12   Stack Sizing for Zone 3

Pipe size, in	Fixture-unit discharge on stack from a branch	Total fixture units allowed on stack
$1\frac{1}{2}$	2	4
2	6	10
3	20*	48*
4	90	240

Note: No more than two toilets may be placed on a 3-in branch, and no more than six toilets may be connected to a 3-in stack.

Figure 18.13   Sizing of Tall Stacks in Zone 3

(Stacks with More than Three Branch Intervals)

Pipe size, in	Fixture-unit discharge on stack from a branch	Total fixture units allowed on stack
1½	2	8
2	6	24
3	20*	72*
4	90	500

Note: No more than two toilets may be placed on a 3-in branch, and no more than six toilets may be connected to a 3-in stack.

To size the stack, first you must determine the fixture load entering the stack at each branch interval. Let's look at an example. We will size a stack that has two branch intervals. The lower branch has a half-bath and a kitchen on it. Using the ratings from zones 2 and 3, we see that the total fixture-unit count for this branch is 6. This is determined by the table listing fixture-unit ratings for various fixtures.

The second stack has a full bathroom group on it. The total fixture-unit count on this branch, using sizing from zones 2 and 3, is 6, if we use a bathroom group rating, or 7, if we count each fixture individually. I would use the larger of the two numbers, for a total of 7.

Next, we look at the sizing table that contains horizontal listings for a 3-in pipe. We know the stack must have a minimum size of 3 in to accommodate the toilets. As we look across the table, we see that each 3-in branch may carry up to 20 fixture units. Well, the first branch has 6 fixture units, and the second branch has 7 fixture units, so both branches are within their limits.

The total from both branches is 13 fixture units. Continuing to look across the table, we see that the stack can accommodate up to 48 fixture units in zone 3 and up to 30 fixture units in zone 2. Obviously, a 3-in stack is adequate. If the fixture-unit loads had exceeded the numbers in either column, the pipe size would have to be increased.

In sizing a stack, the developed length of the stack may comprise different sizes of pipe. For example, at the top of the stack, the pipe size may be 3 in, but at the bottom of the stack the pipe size may be 4 in. The reason is that as you get to the lower portion of the stack, the total number of fixture units placed on the stack is greater.

Sizing Storm Water Drainage Systems

When you wish to size a storm water drainage system (Fig. 18.14), you must have some benchmark information to work with. One consideration is the amount of pitch of a horizontal pipe. Another factor is the number of square feet of surface area that the system will be required to drain. You will also need data on the rainfall rates in the area.

When you use the code book to size a storm water drain system, you should have access to all the key elements required to size the job, except possibly the local rainfall amounts. You can obtain rainfall data from the state or county offices. Your code book contains a table to use in the sizing calculations.

Figure 18.14   Example of a Horizontal Storm Water Sizing Table

Pipe grade, in/ft	Pipe size, in	Gallons per minute (gpm)	Surface area, ft²
$\frac{1}{4}$	3	48	4,640
$\frac{1}{4}$	4	110	10,600
$\frac{1}{4}$	6	314	18,880
$\frac{1}{4}$	8	677	65,200

Note: These figures are based on a rainfall with a maximum rate of 1 in/h of rain, for 1 full hour, and occurring once every 100 years.

Sizing a horizontal storm drain or sewer

The first step in sizing a storm drain or sewer is to list the known data. How much pitch will the pipe have? In this example, my pipe has $\frac{1}{4}$ in/ft pitch.

What else do I know? Well, I know the system is going to be located in Portland, Maine. Portland's rainfall is rated at 2.4 inches per hour (in/h). This rating assumes a 1-h storm that is only likely to occur once every 100 years. Now I have two factors to size the system.

I also know that the surface area to be drained is 15,000 ft^2; this includes the roof and parking area. I've got three of the elements needed to get this job done. But how do I use the numbers in a sizing chart? Well, there are a couple of ways to ease the burden. When you are working with a standard table, like the ones found in most code books, you must adapt the information to suit local conditions. For example, if a standardized table is based on 1 in/h of rainfall and my location has 2.4 in/h of rainfall, I must adjust the table. But this is not difficult; trust me.

When I want to adjust a table based on a 1-in rainfall to meet my local needs, I divide the drainage area in the table by the rainfall amount. For example, if the standard chart shows that an area of 10,000 ft^2 requires a 4-in pipe, I can change the value in the table by dividing the rainfall amount, 2.4, into the surface area, 10,000 ft^2.

Dividing 10,000 by 2.4, I get 4167. All of a sudden, I have solved the mystery of computing storm water piping needs. With this simple conversion, I know that if my surface area is 4167 ft^2, I need a 4-in pipe. But my surface area is 15,000 ft^2, so what size pipe do I need? Well, I know it has to be larger than 4 in. So I look at the conversion chart and find the appropriate surface area. A 15,000-ft^2 surface area requires a storm water drain with a diameter of 8 in.

Now, let's recap this exercise. To size a horizontal storm drain or sewer, decide what pitch will be put on the pipe. Next, determine the area's rainfall for a 1-h storm, occurring each 100 years. If you live in a city, it may be listed, with its rainfall amount, in your code book. Using a standardized chart, rated for 1 in/h of rainfall, divide the surface area by a factor equal to your rainfall index—in my case, it was 2.4. This division process converts a generic table value to a customized table value, just for your area.

Once the math is done, scan the table for the surface area that most closely matches the area you have to drain. To be safe, use a number slightly higher than your projected number. It is better to have a pipe sized one size too large than one size too small. When you have found the appropriate surface area, look across the table to find the pipe size you need. See how easy that was. Well, maybe it's not easy, but it is a chore you can handle.

Sizing rain leaders and gutters

When you need to size rain leaders or downspouts, use the same procedure described above, with one exception. Use a table in the code book to size the *vertical* piping. Determine the amount of surface area your leader will drain, and use the appropriate table to find the pipe size. The conversion factors are the same.

Sizing gutters is essentially the same as sizing horizontal storm drains. You will use a different table in the code book, but the mechanics are the same.

Roof drains

Roof drains are often the starting point of a storm water drainage system. As the name implies, roof drains are located on roofs. On most roofs, the drains are equipped with strainers that protrude upward, at least 4 in, to catch leaves and other debris. Roof drains should be at least twice the size of the piping connected to them. All roofs that do not drain to hanging gutters are required to have roof drains. A minimum of two roof drains should be installed on roofs with a surface area of 10,000 ft^2 or less. If the surface area exceeds 10,000 ft^2, a minimum of four roof drains should be installed.

When a roof is used for purposes other than just shelter, the roof drains may have a strainer that is flush with the roof's surface. Roof drains obviously should be sealed to prevent water from leaking around them. The size of the roof drain is key in the flow rates designed for a storm water system. When a controlled flow from roof drains is wanted, the roof structure must be designed to accommodate it.

More sizing information

If a combined storm drain and sewer arrangement is approved, it must be sized properly. This requires converting fixture-unit loads to drainage surface area. For example, 256 fixture units are treated as 1000 ft^2 of surface area. Each additional fixture unit in excess of 256 is assigned a value of $3\frac{9}{10}$ ft^2. In sizing for continuous flow, each gallon per minute (gpm) is rated as 96 ft^2 of drainage area.

Sizing Vents

Sizing an individual vent is easy. The vent must be at least one-half the size of the drain it serves, but it may not have a diameter of less than $1\frac{1}{4}$ in. For example, a vent for a 3-in drain could, in most cases, have a diameter of $1\frac{1}{2}$ in. A vent for a $1\frac{1}{2}$-in drain may not have a diameter of less than $1\frac{1}{4}$ in (Figs. 18.15 and 18.16).

Relief vents

Relief vents are used in conjunction with other vents. Their purpose is to provide additional air to the drainage system, when the primary vent is too far from the fixture. The relief vent must be at least one-half the size of the pipe it is venting. For example, if a relief vent is venting a 3-in pipe, the relief vent must have a $1\frac{1}{2}$-in, or larger, diameter.

Figure 18.15   Vent Sizing for Zone 3

(For Use with Individual, Branch, and Circuit Vents for Horizontal Drainpipes)

Drainpipe size, in	Drainpipe grade, in/ft	Vent pipe size, in	Maximum developed length of vent pipe, ft
$1\frac{1}{2}$	$\frac{1}{4}$	$1\frac{1}{4}$	Unlimited
$1\frac{1}{2}$	$\frac{1}{4}$	$1\frac{1}{2}$	Unlimited
2	$\frac{1}{4}$	$1\frac{1}{4}$	290
2	$\frac{1}{4}$	$1\frac{1}{2}$	Unlimited
3	$\frac{1}{4}$	$1\frac{1}{2}$	97
3	$\frac{1}{4}$	2	420
3	$\frac{1}{4}$	3	Unlimited
4	$\frac{1}{4}$	2	98
4	$\frac{1}{4}$	3	Unlimited
4	$\frac{1}{4}$	4	Unlimited

Circuit vents

Circuit vents are used with a battery of plumbing fixtures. Circuit vents are normally installed just before the last fixture of the battery. Then, the circuit vent is extended upward to open air or is tied into another vent that extends to the outside. Circuit vents may tie into stack vents or vent stacks. When sizing a circuit vent, you must account for its developed length. But in any event, the diameter of a circuit vent must be at least one-half the size of the drain it is serving.

Vent Sizing Using Developed Length

What effect does the length of the vent have on the vent size? The *developed length*—the total linear footage of pipe making up the vent—is used in conjunction with factors provided in code books to determine vent sizes. To size circuit vents, branch vents, and individual vents for horizontal drains, you must use this method of sizing.

Figure 18.16   Vent Sizing for Zone 3

(For Use with Vent Stacks and Stack Vents)

Drainpipe size, in	Fixture-unit load on drainpipe	Vent pipe size, in	Maximum developed length of vent pipe, ft
$1\frac{1}{2}$	8	$1\frac{1}{4}$	50
$1\frac{1}{2}$	8	$1\frac{1}{2}$	150
$1\frac{1}{2}$	10	$1\frac{1}{4}$	30
$1\frac{1}{2}$	10	$1\frac{1}{2}$	100
2	12	$1\frac{1}{2}$	75
2	12	2	200
2	20	$1\frac{1}{2}$	50
2	20	2	150
3	10	$1\frac{1}{2}$	42
3	10	2	150
3	10	3	1040
3	21	$1\frac{1}{2}$	32
3	21	2	110
3	21	3	810
3	102	$1\frac{1}{2}$	25
3	102	2	86
3	102	3	620
4	43	2	35
4	43	3	250
4	43	4	980
4	540	2	21
4	540	3	150
4	540	4	580

The factors used in sizing a vent, based on developed length, are the grade of the drainage pipe, the size of the drainage pipe, the developed length of the vent, and those allowed by local code requirements. Let's try a few examples of sizing a vent by using this method.

First, assume the drain we are venting is a 3-in pipe with a $\frac{1}{4}$ in/ft grade. This sizing exercise is for zone 3 requirements. If you have a code book for zone 3, inspect the sizing charts in it. Note the number listed under the $1\frac{1}{2}$-in vent column. It is 97. This means that a 3-in drain running horizontally with a $\frac{1}{4}$ in/ft grade can be vented with a $1\frac{1}{2}$-in vent that has a developed length of 97 ft. It would be rare to extend a vent anywhere near 97 ft, but if your vent needed to exceed this distance, you could use a larger vent. A 2-in vent allows you to extend the vent for a total length of 420 ft. A vent larger than 2 in allows you to extend the vent indefinitely.

Second, still in zone 3, assume the drain is a 4-in pipe with a $\frac{1}{4}$ in/ft grade. In this case you could not use a $1\frac{1}{2}$-in vent. Remember, the vent must be at least one-half the size of the drain it is venting. A 2-in vent allows a developed vent length of 98 ft, and a 3-in vent allows the vent to extend to an unlimited length. As you can see, this type of sizing is not difficult.

Now, let's size a vent according to zone 1 rules. In zone 1, vent sizing is based on the vent's length and the number of fixture units on the vent. To size a vent for a lavatory, you need to know how many fixture units the lavatory represents. Lavatories are rated as 1 fixture unit. Using a table in the code book, we find that a vent serving 1 fixture unit can have a diameter of $1\frac{1}{4}$ in and may extend for 45 ft. A bathtub, rated at 2 fixture units, requires a $1\frac{1}{2}$-in vent. The bathtub vent could run for 60 ft.

Branch vents

Branch vents are vents extending horizontally and connecting multiple vents together. They are sized according to the developed-length method, just as in the examples above. A branch vent or individual vent that is the same size as the drain it serves has an unlimited developed length. Be advised, in zones 2 and 3, different tables and ratings are used to size various types of vents; in zone 1, the same rating and table are used for all normal venting situations.

Vent stacks

A *vent stack* is a pipe used only for the purpose of venting. Vent stacks extend upward from the drainage piping to open air, outside a building. Vent stacks are used as connection points for other vents, such as branch vents. A vent stack is a primary vent that accepts the connection of other vents and vents an entire system. Vent stacks run vertically and are sized a little differently.

The basic procedure for sizing a vent stack is similar to that used with branch vents, but there are some differences. You must know the size of the soil stack, the number of fixture units carried by the soil stack, and the developed length of the vent stack. With this information and the regulations of the local plumbing code, you can size your vent stack. Let's work on an example.

Assume the system has a soil stack with a diameter of 4 in. This stack is loaded with 43 fixture units. The vent stack will have a developed length of 50 ft. What size pipe do we use for the vent stack? Looking at the table, we see that a 2-in pipe, used as a vent for the described soil stack, allows a developed length of 35 ft. The vent will have a developed length of 50 ft, so we can rule out 2-in pipe. In the column for $2\frac{1}{2}$-in pipe, the rating is up to 85 ft. Since the vent is going only 50 ft, we could use a $2\frac{1}{2}$-in pipe. However, since $2\frac{1}{2}$-in pipe is not common, we would probably use a 3-in pipe. This same sizing method is used to compute the size of stack vents.

Stack vents

Stack vents are really two pipes in one. The lower portion of the pipe is a soil pipe, and the upper portion is a vent. This is the type of primary vent most often found in residential plumbing. Stack vents are sized by the same methods as those used for vent stacks.

Common vents

Common vents are single vents that vent multiple traps. They are allowed only when the fixtures being served by the single vent are on the same floor. In zone 1, the drains of fixtures being vented with a common vent enter the drainage system at the same level. Normally, not more than two traps can share a common vent, but there is an exception in zone 3. In zone 3, traps of up to three lavatories can be vented with a single common vent. Common vents are sized by the same techniques as those applied to individual vents.

Wet vents

Wet vents (Figs. 18.17 and 18.18) are pipes that serve as a vent for one fixture and a drain for another. Wet vents, once you know how to use them, can save you a lot of money and time. By effectively using wet vents, you can reduce the amount of pipe, fittings, and labor required to vent a bathroom group, or two.

The sizing of wet vents is based on fixture units. The size of the pipe is determined by the number of fixture units it may be required to carry. A 3-in wet vent can handle 12 fixture units. A 2-in wet vent is rated for 4 fixture units, and a $1\frac{1}{2}$-in wet vent is allowed only 1 fixture unit. It is acceptable to wet vent two bathroom groups, six fixtures, with a

Figure 18.17 Sizing a Vent Stack for Wet Vents in Zone 2

Wet-vented fixtures	Vent stack size requirements, in
1–2 Bathtubs or showers	2
3–5 Bathtubs or showers	$2\frac{1}{2}$
6–9 Bathtubs or showers	3
10–16 Bathtubs or showers	4

Figure 18.18   Sizing a Wet Stack Vent in Zone 2

Pipe size of stack, in	Fixture-unit load on stack	Maximum length of stack, ft
2	4	30
3	24	50
4	50	100
6	100	300

single vent, but the bathroom groups must be on the same floor. In zone 2, provisions are made for wet-venting bathrooms on different floors. In zone 1, the approach to wet vents is different.

In zone 2 there are additional regulations that pertain to wet vents: The horizontal branch connecting to the drainage stack must enter at a level equal to, or below, the water closet drain. However, the branch may connect to the drainage at the closet bend. When there are two bathroom groups, the wet vent must have a minimum diameter of 2 in.

Kitchen sinks and washing machines may not be drained into a 2-in combination waste-and-vent. Water closets and urinals are restricted on vertical combination waste-and-vent systems.

As for the permissibility allowance of wet venting on different levels in zone 2, here are the facts. Wet vents must have at least a 2-in diameter. Water closets that are not located on the highest floor must be back-vented. If, however, the wet vent is connected directly to the closet bend with a 45° bend, the toilet being connected is not required to be back-vented, even if it is on a lower floor.

In zone 1 wet vents are limited to vertical piping. These vertical pipes are restricted to receiving only the waste from fixtures with fixture-unit ratings of 2 or less and that vent no more than four fixtures. Wet vents must be one pipe size larger than normally required, but they must never be smaller than 2 in in diameter.

Crown vents

A *crown vent* is a vent that extends upward from a trap or trap arm. Crown vent traps are not allowed. Crown vents are normally used on trap arms, but even then, they are not common. The vent must be on the trap arm, and it must be behind the trap by a distance equal to twice the pipe size. For example, on a $1\frac{1}{2}$-in trap, the crown vent has to be 3 in behind the trap, on the trap arm.

Vents for Sumps and Sewer Pumps

When sumps and sewer pumps are used to store and remove sanitary waste, the sump must be vented. In zones 1 and 2, these vents are treated about the same as vents installed on gravity systems.

If you will be installing a pneumatic sewer ejector, you need to run the sump vent to outside air, without tying it into the vents for the standard sanitary plumbing system. This ruling on pneumatic pumps applies to all three zones. If the sump will be equipped with a regular sewer pump, you may tie the vent from the sump back into the main venting system for the other sanitary plumbing.

In zone 3 there are some additional rules. The following is an outline of the requirements in zone 3: Sump vents may not be smaller than $1\frac{1}{4}$-in pipe. The size requirements for sump vents are determined by the discharge of the pump. For example, a sewer pump capable of producing 20 gpm could have its sump vented for an unlimited distance with $1\frac{1}{2}$-in pipe. If the pump is capable of producing 60 gpm, a $1\frac{1}{2}$-in pipe could not have a developed length of more than 75 ft.

In most cases, a 2-in vent is used on sumps, and the distance allowed for developed length is not a problem. However, if the pump will pump more than 100 gpm, you had better take the time to do some math. The code book has the factors you need to size the vent, and the sizing is easy. Simply look for the maximum discharge capacity of

Figure 18.19   Trap-to-Vent Distances in Zone 1

Grade on drainpipe, in	Size of trap arm, in	Maximum distance between trap and vent
$\frac{1}{4}$	$1\frac{1}{4}$	2 ft 6 in
$\frac{1}{4}$	$1\frac{1}{2}$	3 ft 6 in
$\frac{1}{4}$	2	5 ft
$\frac{1}{4}$	3	6 ft
$\frac{1}{4}$	4 and larger	10 ft

Figure 18.20   Trap-to-Vent Distances in Zone 2

Grade on drainpipe, in	Fixture's drain size, in	Trap size, in	Maximum distance between trap and vent
$\frac{1}{4}$	$1\frac{1}{4}$	$1\frac{1}{4}$	3 ft 6 in
$\frac{1}{4}$	$1\frac{1}{2}$	$1\frac{1}{4}$	5 ft
$\frac{1}{4}$	$1\frac{1}{2}$	$1\frac{1}{2}$	5 ft
$\frac{1}{4}$	2	$1\frac{1}{2}$	8 ft
$\frac{1}{4}$	2	2	6 ft
$\frac{1}{8}$	3	3	10 ft
$\frac{1}{8}$	4	4	12 ft

the pump, and match it with a vent that allows the developed length you need.

Another consideration in sizing vents is the distance between the vent and the trap it is serving. The grade of a drainpipe can also influence the size of a system. Again, information in your local code book (Figs. 18.19, 18.20, and 18.21) will help you determine the proper pipe size.

Sizing Potable Water Systems

In this section we show how to size potable water piping. But, be advised, this procedure is not always simple, and it requires concentration. Some parts of potable water pipe sizing are not very hard. Many times, the code book con-

Figure 18.21 Trap-to-Vent Distances in Zone 3

Grade on drainpipe, in	Fixture's drain size, in	Trap size, in	Maximum distance between trap and vent
$\frac{1}{4}$	$1\frac{1}{4}$	$1\frac{1}{4}$	3 ft 6 in
$\frac{1}{4}$	$1\frac{1}{2}$	$1\frac{1}{4}$	5 ft
$\frac{1}{4}$	$1\frac{1}{2}$	$1\frac{1}{2}$	5 ft
$\frac{1}{4}$	2	$1\frac{1}{2}$	8 ft
$\frac{1}{4}$	2	2	6 ft
$\frac{1}{8}$	3	3	10 ft
$\frac{1}{8}$	4	4	12 ft

Figure 18.22 Recommended Minimum Sizes for Fixture Supply Pipes

Fixture	Minimum pipe size, in
Bathtub	$\frac{1}{2}$
Bidet	$\frac{3}{8}$
Dishwasher	$\frac{1}{2}$
Hose bib	$\frac{1}{2}$
Kitchen sink	$\frac{1}{2}$
Laundry tub	$\frac{1}{2}$
Lavatory	$\frac{3}{8}$
Shower	$\frac{1}{2}$
Water closet, two-piece	$\frac{3}{8}$
Water closet, one-piece	$\frac{1}{2}$

tains charts and tables (Fig. 18.22) to help you. Some of these graphics will detail precisely what size of pipe or tubing is required. But, unfortunately, code books cannot provide concrete answers for all piping installations.

Many factors affect the sizing of potable water piping. The type of pipe used affects your findings. Some pipe materials have smaller inside diameters than others, and other pipe materials have a rougher surface or more restrictive fittings than others. Both factors affect the sizing of a water system.

When sizing a potable water system, you must be concerned with the speed of the flowing water, the quantity of water (Figs. 18.23 and 18.24) needed, and the restrictive qualities of the pipe being used to convey the water. Most materials approved for potable water piping will provide a flow velocity of 5 feet per second (ft/s). The exception is galvanized-steel pipe. Galvanized-steel pipe provides a flow speed of 8 ft/s.

It may surprise you that galvanized-steel pipe has a faster flow rate. This is the result of the wall strength of galvanized-steel pipe. In softer pipes, such as copper, fast-moving water can essentially wear a hole in the pipe. These flow ratings are not etched in stone. Undoubtedly some people will argue for either a higher or a lower rating, but these ratings are used in current plumbing codes.

Figure 18.23 Recommended Capacities at Fixture Supply Outlets

Fixture	Flow rate, gpm	Flow pressure, lb/in^2
Bathtub	4	8
Bidet	2	4
Dishwasher	2.75	8
Hose bib	5	8
Kitchen sink	2.5	8
Laundry tub	4	8
Lavatory	2	8
Shower	3	8
Water closet, two-piece	3	8
Water closet, one-piece	6	20

Figure 18.24 Sample Pressure and Pipe Chart for Sizing

Size of water meter and street service, in	Size of water service and distribution pipes, in	Maximum length of water pipe, ft					
		40	60	80	100	150	200
$3/4$	$1/2$	9	8	7	6	5	4
$3/4$	$3/4$	27	23	19	17	14	11
$3/4$	1	44	40	36	33	28	23
1	1	60	47	41	36	30	25
1	$1\frac{1}{4}$	102	87	76	67	52	44

To use the three factors previously discussed to determine the pipe size, you must use math that you may not have seen since schooldays, and you may not have seen it then. Let me give you an example of how a typical formula might look. A common formula might look like this:

$$Y = XZ$$

where X = flow rate of water, 5 ft/s usually
Y = amount of water in pipe
Z = inside diameter of pipe

Since many plumbers refuse to do this type of math, most code books offer alternatives. The alternatives are often tables or charts that show pertinent information on the requirements for pipe sizing.

The tables or charts in a code book are likely to discuss the following: outside diameter of the pipe, inside diameter of the pipe, flow rate for the pipe, and pressure loss in the pipe over a distance of 100 ft. These charts or tables are specific to a particular type of pipe. For example, there is one table for copper pipe and another table for PB pipe.

The information supplied in a ratings table for PB pipe might look like this:

- Pipe size is $\frac{3}{4}$ in.

- Inside pipe diameter is 0.715.

- The flow rate, at 5 ft/s, is 6.26 gpm.

- Pressure lost in 100 ft of pipe is 14.98.

This type of pipe sizing is most often done by engineers, not plumbers. When you size a potable water system, start at the last fixture and work your way back to the water service.

Commercial jobs, where pipe sizing can get quite complicated, are generally sized by design experts. All that a working plumber is required to do is to install the proper pipe sizes in the proper locations and manner. For residential plumbing, where engineers are less likely to have a hand in the design, there is a rule-of-thumb method to sizing most jobs. In the average home, a $\frac{3}{4}$-in pipe is sufficient for the main artery of the water distribution system. Normally, not more than two fixtures can be served by a $\frac{1}{2}$-in pipe. With this in mind, sizing becomes simple.

Usually $\frac{3}{4}$-in pipe is run to the water heater, and it is typically used as a main water distribution pipe. When it nears the end of a run, the $\frac{3}{4}$-in pipe is reduced to $\frac{1}{2}$-in pipe, once there are no more than two fixtures to connect to. Most water service pipes will have a $\frac{3}{4}$-in diameter, with those serving homes with numerous fixtures having a 1-in diameter. This rule-of-thumb sizing will work on almost any single-family residence.

The water supplies to fixtures must meet minimum standards. These sizes are derived from local code requirements. Simply find the fixture you are sizing the supply for, and check the column heading for the proper size.

Most code requirements seem to agree that there is no definitive way to set a boilerplate formula for establishing potable water pipe sizing. Code officers can require pipe sizing to be performed by a licensed engineer. In most major plumbing systems, the pipe sizing is done by a design professional.

Code books give examples of how a system might be sized. But the examples are not meant as a code requirement. The code requires a water system to be sized properly. However, due to the complexity of the process, the books do not set firm statistics for the process. Instead, code books give parts of the puzzle, in the form of some minimum standards, but it is up to a professional designer to come up with an approved system.

Where does this leave you? Well, the sizing of a potable water system is one of the most complicated aspects of plumbing. Very few single-family homes are equipped with potable water systems designed by engineers. I have already given you a basic rule-of-thumb method for sizing small systems. Next, I will show you how to use the fixture-unit method of sizing.

The Fixture-Unit Method

The fixture-unit method is simple, and it is generally acceptable to code officers. While this method may not be perfect, it is much faster and easier to use than the velocity method. Except for the additional expense for materials, you can't go wrong by oversizing pipe. If you are in doubt about the size, go to the next larger size. Now, let's see how to size a single-family residence's potable water system by using the fixture-unit method.

Most codes assign a fixture-unit value (Fig. 18.25) to common plumbing fixtures. To size by the fixture-unit method, you must establish the number of fixture units to be carried by the pipe. You must also know the working pressure of the water system. Most codes provide guidelines for these two pieces of information.

For this example, the house has the following fixtures: three toilets, three lavatories, one bathtub/shower combination, one shower, one dishwasher, one kitchen sink, one laundry hookup, and two sill cocks. The water pressure serving this house is 50 psi. There is a 1-in water meter serving the house, and the water service is 60 ft long. With this information and the guidelines provided by the local code, we can do a pretty fair job of sizing the potable water system.

Figure 18.25 Common Fixture-Unit Values for Water Distribution

Fixture	Hot	Cold	Total
Bathtub	3	6	8
Bidet	1.5	1.5	2
Kitchen sink	1.5	1.5	2
Laundry tub	2	2	3
Lavatory	1.5	1.5	2
Shower	3	3	4
Water closet, two-piece	0	5	5

The first step is to establish the total number of fixture units on the system. The code regulations provide this information. There are three toilets, so that's 9 fixture units. The three lavatories add 3 fixture units. The tub/shower combination counts as 2 fixture units, and the showerhead over the bathtub doesn't count as an additional fixture. The shower has 2 fixture units. The dishwasher adds 2 fixture units, and so does the kitchen sink. The laundry hookup counts as 2 fixture units. Each sill cock is worth 3 fixture units. This house has a total fixture-unit load of 28.

Now we have the first piece of the sizing puzzle. The next step is to determine what size pipe will allow the number of fixture units. This house has a water pressure of 50 psi. This pressure rating falls into the category allowed in the sizing chart in the code book. First, we find the proper water meter size—1-in. Note that a 1-in meter and a 1-in water service are capable of handling 60 fixture units, when the pipe is only running 40 ft. However, when the pipe stretches to 80 ft, the fixture-unit load is reduced to 41. At 200 ft, the fixture-unit rating is 25. What is it at 100 ft? At 100 ft, the allowable fixture load is 36. See, this type of sizing is not so hard.

Now, what does this tell us? Well, we know the water service is 60 ft long. Once inside the house, how far is it to the most remote fixture? In this case, the farthest fixture is 40 ft from the water service entrance. This gives a developed length of 100 ft—60 ft for the water service and 40 ft for the interior pipe. Going back to the sizing table, we see that for 100 ft of pipe, under the conditions in this example, we are allowed 36 fixture units. The house has only 28 fixture units, so the pipe is properly sized.

What if the water meter were a ³⁄₄-in meter, instead of a 1-in meter? With a ³⁄₄-in meter and a 1-in water service and main distribution pipe, we could have 33 fixture units. This would still be a suitable arrangement, since we have only 28 fixture units. Could we use a ³⁄₄-in water service and water distribution pipe with the ³⁄₄-in meter? No. With all sizes set at $\frac{3}{4}$ in, the maximum number of fixture units allowed is 17.

In this example, the piping is oversized. But if you want to be safe, follow this type of procedure. If you are required to provide a riser diagram showing the minimum pipe sizing allowed, you will have to do a little more work. Once inside a building, water distribution pipes normally extend for some distance, supplying many fixtures with water. As the distribution pipe continues on its journey, the fixture load is reduced.

For example, assume that the distribution pipe serves a full bathroom group within 10 ft of the water service. Once this group is served with water, the fixture-unit load on the remainder of the water distribution piping is reduced by 6 fixture units. As the pipe serves other fixtures, the fixture-unit load continues to decrease. So it is feasible for the water distribution pipe to become smaller as it goes along.

Let's look at that same house and see how we could use smaller pipe. Okay, we know we need a 1-in water service. Once inside the foundation, the water service becomes the water distribution pipe. The water heater is located 5 ft from the cold water distribution pipe. The 1-in pipe extends over the water heater and supplies it with cold water. And the hot water distribution pipe originates at the water heater. Now we have two water distribution pipes to size.

When sizing the hot and cold water pipes, we could make adjustments for fixture-unit values on some fixtures. For example, a bathtub is rated as 2 fixture units. Since the bathtub rating is inclusive of both hot and cold water, obviously the demand for just the cold water pipe is less than that shown in the table. For simplicity's sake, we do not break down the fixture units to fractions or reduced amounts. We will do the example as if a bathtub required 2 fixture units of hot water and 2 fixture units of cold water. However, we could reduce the amounts listed in the table by about 25 percent to obtain the rating for each hot and cold water pipe. For example, the bathtub, when being sized for only cold water, could take on a value of $1\frac{1}{2}$ fixture units.

Now then, let's get on with the example. We are at the water heater. We run a 1-in cold water pipe overhead and drop a $\frac{3}{4}$-in pipe into the water heater. What size pipe do we bring up for the hot water? First, we count the number of fixtures that use hot water, and we assign them a fixture-unit value. The fixtures using hot water are all fixtures, except the toilets and sill cocks. The total count for hot water fixture units is 13. From the water heater, the most remote hot water fixture is 33 ft away.

What size pipe should we bring up from the water heater? If you look at the sizing table in your code book, you will find a distance and fixture-unit count that will work in this case. Look under the 40-ft column, since the distance is less than 40 ft. The first fixture-unit number in the column is 9; this won't work. The next number is 27; this will work, because it is greater than the 13 fixture units we need. Looking across the table, you will see that the minimum pipe size to start with is $\frac{3}{4}$-in pipe. Isn't it convenient that the water heater just happens to be sized for $\frac{3}{4}$-in pipe?

Okay, now we start our hot water run with $\frac{3}{4}$-in pipe. As our hot water pipe travels along the 33-ft stretch, it provides water to various fixtures. When the total fixture-unit count remaining to be served drops to less than 9, we can reduce the pipe to $\frac{1}{2}$-in pipe. We can also run the fixture branches off the main in $\frac{1}{2}$-in pipe. We can do this because the highest fixture-unit rating on any of the hot water fixtures is 2. Even with a pipe run of 200 ft we can use $\frac{1}{2}$-in pipe for up to 4 fixture units. Is this sizing starting to seem easy? Remember the rule-of-thumb sizing I gave earlier. These sizing examples show how well the rule-of-thumb method works.

With the hot water sizing done, let's look at the remainder of the cold water. There is a pipe run of less than 40 ft to the farthest cold water fixture. There is a branch near the water heater drop for a sill cock, and there is a full bathroom group within 7 ft of the water heater drop. The sill cock branch can be ½-in pipe. The pipe going under the full bathroom group could probably be reduced to ¾-in pipe, but it would be best to run it as a 1-in pipe. However, after the bathroom group and the sill cock have been served, how many fixture units are left? There are only 19. We can now reduce to ¾-in pipe, and when the demand drops to below 9 fixture units, we can reduce to ½-in pipe. All the fixture branches can be run with ½-in pipe.

This is one way to size a potable water system that works without driving you crazy. Some may dispute the sizes given in these examples, saying that the pipe is oversized. But as I said earlier, when in doubt, go bigger. In reality, the cold water pipe in the last example could probably have been reduced to ¾-in pipe where the transition was made from water service to water distribution pipe. It could have almost certainly been reduced to ¾-in pipe after the water heater drop. Local codes will have their own interpretations of pipe sizing, but this method will normally serve you well.

Well, we're done with the sizing exercises. Now it's time to discuss the proper way to design a plumbing system once it has been sized.

19

Design Criteria

What are the design criteria for plumbing systems? The plumbing code is certainly a big part of the criteria. Any system being designed must conform to applicable code requirements. We've already talked about pipe sizing, which is part of designing a system. In this chapter we focus on basic design rules.

We've talked about why commercial plumbers are not normally required to design their own layouts. Residential plumbers, however, are often required to create their own layouts. The designer must keep many factors in mind. The cost of material is one consideration. Efficient designs use minimal amounts of material to create the highest profit possible. Another consideration in layout is the amount of effort required to install it. If you can reduce the labor required to make an installation, you can make more money.

What else should you keep in mind when working out a design? Code requirements must be met. Customer satisfaction is also a factor to consider. Just making your work easier is reason enough to put some thought into making a good design. With this in mind, let's talk about some design issues.

Grading Pipe

When you install horizontal drainage piping, you must install it so that it falls toward the waste disposal site. A typical grade for drainage pipe is a fall of $\frac{1}{4}$ in/ft. This means the lower end of a 20-ft pipe is 5 in lower than the upper end, when properly installed. While the $\frac{1}{4}$ in/ft grade is typical, it is not the only acceptable grade for all pipes.

If you are working with pipe that has a diameter of $2\frac{1}{2}$ in or less, the minimum grade for the pipe is $\frac{1}{4}$ in/ft. Pipes with diameters between 3 and 6 in are allowed a minimum grade of $\frac{1}{8}$ in/ft. In zone 1, special permission must be granted prior to installation of pipe with a $\frac{1}{8}$ in/ft grade. In zone 3, for pipes with diameters of 8 in or more, an acceptable grade is $\frac{1}{16}$ in/ft.

Supporting Pipe

Pipe support is also regulated by the plumbing code (Figs. 19.1, 19.2, and 19.3). There are requirements about the types of materials you may use and how. One concern with the type of hangers used is their compatibility with the pipe they are supporting. You must use a hanger that will not harm the piping. For example, you may not use a gal-

Figure 19.1 Support Intervals for Water Pipe in Zone 1

Type of pipe, in	Vertical support interval	Horizontal support interval, ft
Threaded pipe ($\frac{3}{4}$-in and smaller)	Every other story	10
Threaded pipe (1-in and larger)	Every other story	12
Copper tube ($1\frac{1}{2}$-in and smaller)	Every story, not to exceed 10 ft	6
Copper tube (2-in and larger)	Every story, not to exceed 10 ft	10
Plastic pipe	Not mentioned	4

Figure 19.2 Support Intervals for Water Pipe in Zone 2

Type of pipe	Vertical support interval	Horizontal support interval, ft
Threaded pipe	30 ft	12
Copper tube ($1\frac{1}{4}$ in and smaller)	4 ft	6
Copper tube ($1\frac{1}{2}$ in)	Every story	6
Copper tube (larger than $1\frac{1}{2}$ in)	Every story	10
Plastic pipe (2 in and larger)	Every story	4
Plastic pipe ($1\frac{1}{2}$ in and smaller)	4 ft	4

Figure 19.3 Pipe Support Intervals in Zone 3

Type of vent pipe	Maximum distance between supports
	Horizontal
PB	32 in
Lead	Continuous
Cast iron	5 ft or at each joint
Galvanized steel	12 ft
Copper tube ($1\frac{1}{4}$ in)	6 ft
Copper tube ($1\frac{1}{2}$ ko,in and larger)	10 ft
ABS	4 ft
PVC	4 ft
Brass	10 ft
Aluminum	10 ft
	Vertical
Lead	4 ft
Cast iron	15 ft
Galvanized steel	15 ft
Copper tubing	10 ft
ABS	4 ft
PVC	4 ft
Brass	10 ft
PB	4 ft
Aluminum	15 ft

vanized-steel strap hanger to support copper pipe. As a rule of thumb, the hangers used to support a pipe should be made of the same material as the pipe. For example, copper pipe should be hung with copper hangers. This eliminates the risk of corrosive action between two different types of materials. A plastic or plastic-coated hanger can be used with all types of pipe. An exception to this rule might be when the piping is carrying a liquid with a temperature that might affect or melt the plastic hanger.

The hangers used to support pipe must be capable of supporting the pipe at all times. Hangers must be attached to the pipe and to the member holding the hanger in a satisfactory manner. For example, it is not acceptable to wrap a piece of wire around a pipe and then wrap the wire around the bridging between two floor joists. Hangers should be securely attached to the member supporting it. For example, a hanger should be attached to the pipe and then nailed to a floor joist. The nails used to hold a hanger in place should be made of the same material as the hanger, if corrosive action is a possibility.

Both horizontal and vertical pipes require support. The intervals between supports will vary, depending upon the type of pipe being used and whether it is installed vertically or horizontally. The following examples will show you how often to support various types of pipes when they are hung horizontally. These are examples of the maximum distances allowed between supports for zone 3:

ABS—every 4 ft	Cast iron—every 5 ft
Galvanized steel—every 12 ft	PVC—every 4 ft
DWV copper—every 10 ft	

When these same types of pipes are installed vertically, in zone 3, they must be supported at no less than the following intervals:

ABS—every 4 ft	Cast iron—every 15 ft
Galvanized steel—every 15 ft	PVC—every 4 ft
DWV copper—every 10 ft	

When cast-iron stacks are installed, the base of each stack must be supported. Pipes with flexible couplings, bands, or unions must be installed and supported so as to prevent the flexible connections from moving. In larger pipes—pipes larger than 4 in—all flexible couplings must be supported to prevent the force of the pipe's flow from loosening the connection at changes in direction.

Pipe Size Reduction

You may not reduce the size of a drainage pipe as it heads for the waste disposal site. The pipe size may be enlarged, but it may not be reduced. There is one exception to this rule. Reducing closet bends, such as a 4-by-3 closet bend, is allowed.

Underground Pipes

A drainage pipe installed underground must have a minimum diameter of 2 in. When you are installing a horizontal branch fitting near the base of a stack, keep the branch fitting away from the point where the vertical stack turns to a horizontal run. The branch fitting should be installed at least 30 in back on a 3-in pipe and 40 in back on a 4-in pipe. Multiply the pipe size by 10 to determine how far back the branch fitting should be installed.

All drainage piping must be protected from the effects of flooding. When a stub of pipe is left to connect with fixtures planned for the future, the stub must not be more than 2 ft long, and it must be capped. Some exceptions are possible on the prescribed length of a pipe stub. If you need a longer stub, consult the local code officer. Cleanout extensions are not affected by the 2-ft rule.

Fittings

Fittings are also a part of the drainage system. Knowing when, where (Fig. 19.4), and how to use the proper fittings is essential to the installation of a drainage system. Fittings are used to make branches and to change direction. The use of fittings to change direction is where we will

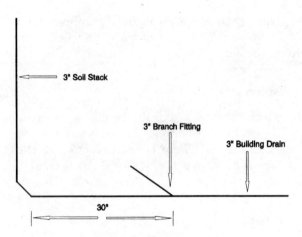

Figure 19.4 Branch-fitting rule.

start. When you wish to change direction with pipe, you
may change it from a horizontal run to a vertical rise. You
may be going from a vertical position to a horizontal one, or
you might only want to offset the pipe in a horizontal run.
Each of these three categories requires the use of different
fittings. Let's take each circumstance and examine the fittings allowed.

Offsets in horizontal piping

When you want to change the direction of a horizontal pipe,
you must use fittings approved for that purpose. You have
six choices in zone 3:

Sixteenth bend	Eighth bend
Sixth bend	Long-sweep fittings
Combination wye and eighth bend	Wye

Any of these fittings is generally approved for changing
direction with horizontal piping; but as always, it is best to
check with your local code officer for current regulations.

Horizontal to vertical

There is a wider choice in fittings for going from a horizontal position to a vertical position. There are nine possible candidates in zone 3:

Sixteenth bend	Eighth bend
Sixth bend	Long-sweep fittings
Combination wye and eighth bend	Wye
Quarter bend	Short-sweep fittings
Sanitary tee	

You may not use a double sanitary tee in a back-to-back situation if the fixtures being served are of a blowout or pump type. For example, you could not use a double sanitary tee to receive the discharge of two washing machines, if the machines were positioned back to back. The sanitary tee's throat is not deep enough to keep drainage from feeding back and forth between the fittings. In a case like this, use a double combination wye and eighth bend. The combination fitting has a much longer throat and will prohibit waste water from transferring across the fitting to the other fixture.

Vertical to horizontal

Seven fittings can be used to change direction from vertical to horizontal:

Sixteenth bend	Eighth bend
Sixth bend	Long-sweep fittings
Combination wye and eighth bend	Wye
Short-sweep fittings 3-in or larger	

Indirect Wastes

Indirect-waste requirements can pertain to a number of types of plumbing fixtures and equipment. These might include a clothes washer drain, a condensate line, a sink drain, or the blow-off pipe from a relief valve, just to name a few. These indirect wastes are piped in this manner to prevent the possibility of contaminated matter backing up the drain into a potable water or food source, among other things.

Most indirect-waste receptors are trapped. If the drain from the fixture is more than 2 ft long, the indirect-waste receptor must have a trap. However, this rule applies to fixtures like sinks, not to an item such as a blow-off pipe from a relief valve. The rule is different in zone 1. In zone 1, if the drain is more than 5 ft long, it must have a trap.

The safest disposal method for an indirect waste is accomplished by using an air gap. When an air gap is used, the drain from the fixture terminates above the indirect-waste receptor, with open airspace between the waste receptor and the drain. This prevents any backup or back-siphonage.

Some fixtures, depending on local code requirements, may be piped with an air break, rather than an air gap. With an air break, the drain may extend below the flood-level rim and terminate just above the trap's seal. The risk with an air break is the possibility of a backup. Since the drain is run below the flood-level rim of the waste receptor, the waste receptor could overflow and back up into the drain. This could create contamination, but in cases where contamination is likely, an air gap will be required. Check with the local code officer before you use an air break.

Standpipes, like those used for washing machines, are a form of indirect-waste receptor. A standpipe (Fig. 19.5) used for this purpose in zones 1 and 3 must extend at least 18 in above the trap's seal, but it may not extend more than 30 in above the trap's seal. If a clear-water waste receptor is located in a floor, in zone 3 the lip of the receptor must extend at least 2 in above the floor. This eliminates the waste receptor from being used as a floor drain.

Buildings used for food preparation, storage, and similar activities are required to have their fixtures and equipment discharge drainage through an air gap. In zone 3 there is an exception to this rule. In zone 3, dishwashers and open culinary sinks are excepted. In zone 2 the discharge pipe

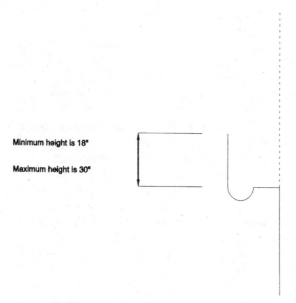

Minimum height is 18"

Maximum height is 30"

Figure 19.5 Washing machine receptor rule.

must terminate at least 2 in above the receptor. In zone 1 the distance must be a minimum of 1 in. In zones 2 and 3 the air-gap distance must be a minimum of twice the size of the pipe discharging the waste. For example, a $\frac{1}{2}$-in discharge pipe requires a 1-in air gap.

In zones 2 and 3 the installation of an indirect-waste receptor is prohibited in any room containing toilet facilities. The same holds in zone 1 but there is one exception. The exception is the installation of a receptor for a clothes washer, when the clothes washer is installed in the same room. Indirect-waste receptors may not be allowed to be installed in closets and other unvented areas. Indirect-waste receptors must be accessible. In zone 2 all receptors must be equipped with a means of preventing solids with diameters of $\frac{1}{2}$ in or larger from entering the drainage system. These straining devices must be removable, to allow for cleaning.

When you are dealing with extreme water temperatures in waste water, such as with a commercial dishwasher, the dishwasher drain must be piped to an indirect waste. The indirect waste will be connected to the sanitary plumbing system, but the dishwasher drain may not connect to the sanitary system directly, if the waste water temperature exceeds 140°F. Steam pipes may not be connected directly to a sanitary drainage system. Local regulations may require the use of special piping, sumps, or condensers to accept high-temperature water. In zone 1 direct connection of any dishwasher to the sanitary drainage system is prohibited.

Clear-water waste, from a potable source, must be piped to an indirect-waste receptor with the use of an air gap. Sterilizers and swimming pools are two examples of when this rule applies. Clear water from nonpotable sources, such as a drip from a piece of equipment, must be piped to an indirect waste. In zone 3, an air break is allowed in place of an air gap. In zone 2 any waste entering the sanitary drainage system from an air conditioner must do so through an indirect waste.

Special Wastes

Special wastes are those wastes that may be harmful to a plumbing system or the waste disposal system. Possible locations for special-waste piping might include photographic labs, hospitals, or buildings where chemicals or other potentially dangerous wastes are dispersed. Small, personal-type photograhic darkrooms do not generally fall under the scrutiny of these regulations. Buildings that are considered to have a need for special-wastes plumbing are often required to have two plumbing systems, one system for normal sanitary discharge and a separate system for the special wastes. Before many special wastes are allowed to enter a sanitary drainage system, the wastes must be neutralized, diluted, or otherwise treated.

Depending upon the nature of the special wastes, special materials may be required. When you venture into the plumbing of special wastes, it is always best to consult the local code officer before proceeding with your work.

Vent Installation Requirements

Since there are so many types of vents (Fig. 19.6) and their role in the plumbing system is so important, there are many regulations affecting their installation. What follows are the specifics for installing various vents.

In zone 2, any building equipped with plumbing must also be equipped with a main vent (Fig. 19.7). In zone 3 any plumbing system that receives the discharge from a water closet must have either a main vent stack or a stack vent. This vent must originate at a 3-in drain pipe and extend upward until it penetrates the roof of the building and meets outside air. The vent size requirements for both zones 2 and 3 call for a minimum diameter of 3 in. However, in zone 2, the main stack in detached buildings, where the only plumbing is a washing machine or laundry tub, may have a diameter of $1\frac{1}{2}$ in. In zone 1, all plumbing fixtures must be vented.

When a vent penetrates a roof, it must be flashed or sealed to prevent water from leaking past the pipe and through the roof. Metal flashings with rubber collars are normally used for flashing vents, but more modern flashings are made of plastic, rather than metal.

The vent must extend above the roof to a certain height. The height may vary from one geographical location to another. Average vent extensions are between 12 and 24 in, but check the local regulations to determine the minimum height in your area. In zones 1 and 2, generally height requirements for vent terminations are set at 6 in above the roof. In zone 3 the vent must extend at least 12 in above the roof.

When vents terminate in open air, the proximity of their location to windows, doors, or other ventilating openings must be considered. If a vent were placed too close to a window, sewer gas could be drawn into the building when the

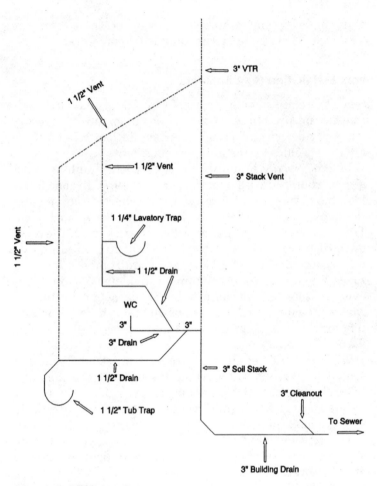

Figure 19.6 DWV riser diagram.

window was open. Vents should be kept 10 ft from any window, door, opening, or ventilation device. If the vent cannot be kept at least 10 ft from the opening, the vent should extend at least 3 ft above the opening. In zone 1, these vents must extend at least 3 ft above the opening.

If the roof being penetrated by a vent is used for activi-

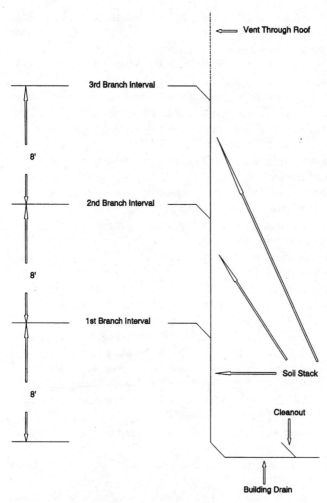

Figure 19.7 Branch-interval detail.

ties other than just weather protection, such as a patio, the vent must extend 7 ft above the roof in zone 3. In zone 2 these vents must rise at least 5 ft above the roof. In cold climates, vents must be protected from freezing. Condensation can collect on the inside of vent pipes. In cold climates this condensation may turn to ice. As the ice mass grows, the vent becomes blocked and useless.

This type of protection is usually afforded by increasing the size of the vent pipe. This rule normally applies only in areas where temperatures are expected to be below 0°F. In zone 3, vents in this category must have a minimum diameter of 3 in. If this requires an increase in pipe size, the increase must be made at least 1 ft below the roof. In the case of sidewall vents, the change must be made at least 1 ft inside the wall.

In zone 1, the rules for protecting vents from frost and snow are a little different. All vents must have diameters of at least 2 in, but never less than the normally required vent size. Any change in pipe size must occur at least 12 in before the vent penetrates open air, and the vent must extend to a height of 10 in.

On some occasions it is better to terminate a plumbing vent out the side of a wall than through a roof. In zone 1 sidewall venting is prohibited. In zone 2, sidewall vents may not terminate under any building's overhang. When sidewall vents are installed, they must be protected against birds and rodents by a wire mesh or similar cover. Sidewall vents must not extend closer than 10 ft to the property boundary of the building lot. If the building is equipped with soffit vents, sidewall vents may not be used in such a way that they terminate under the soffit vents. This rule is, in effect, to prevent sewer gas from being sucked into the attic of the home.

In zone 3, buildings having soil stacks with more than five branch intervals (Fig. 19.8) must be equipped with a vent stack. In zone 1, a vent stack is required with buildings having at least 10 stories above the building drain. The vent stack normally runs up near the soil stack. The vent stack must connect to the building drain at or below the lowest branch interval. The vent stack must be sized according to the instructions given earlier. In zone 3, the vent stack must be connected within 10 times its pipe size on the downward side of the soil stack. This means that a

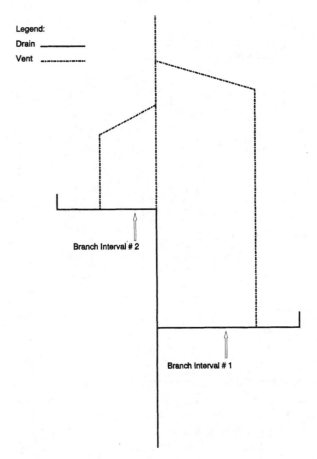

Figure 19.8 Stack with two branch intervals.

3-in vent stack must be within 30 in of the soil stack, on the downward side of the building drain.

In zone 1, these stack vents also must be connected to the drainage stack at intervals of every five stories. The connection must be made with a relief yoke vent. The yoke vent (Fig. 19.9) must be at least as large as either the vent stack or the soil stack, whichever is smaller. This connection must be made with a wye fitting, at least 42 in off the floor.

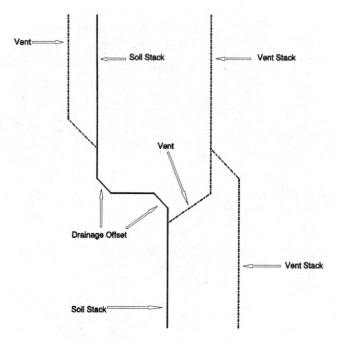

Figure 19.9 Example of venting drainage offsets.

In large plumbing jobs where there are numerous branch intervals, it may be necessary to vent offsets in the soil stack. Normally, the offset must be more than 45° to warrant an offset vent. In zones 2 and 3, offset vents are required when the soil stack offsets and has five, or more, branch intervals above it.

Just as drains are installed with a downward pitch, vents must be installed with a consistent grade. Vents should be graded to allow any water entering the vent pipe to drain into the drainage system. A typical grade for vent piping is a $\frac{1}{4}$ in/ft. In zone 1, vent pipes may be installed level, without pitch.

Dry vents must be installed in a manner to prevent clogging and blockages. You may not lay a fitting on its side and use a quarter bend to turn the vent up vertically. Dry

Figure 19.10   Stack-Venting without Individual Vents in Zone 2

Fixtures allowed to be stack-vented without individual vents*
Water closets
Basins
Bathtubs
Showers
Kitchen sinks, with or without dishwasher and garbage disposal

Note: Restrictions apply to this type of installation.

vents should leave the drain pipe in a vertical position. An easy way to remember this is that if you need an elbow to get the vent up from the drainage, you are doing it wrong.

Most vents can be tied into other vents, such as a vent stack or stack vent (Fig. 19.10). But the connection for the tie-in must be at least 6 in above the flood-level rim of the highest fixture served by the vent.

In zone 2, circuit vents can be used to vent fixtures in a battery. The drain serving the battery must be operating at one-half its fixture-unit rating. If the application is on a lower-floor battery with a minimum of three fixtures, relief vents are required. You must also pay attention to the fixtures draining above these lower-floor batteries.

When a fixture with a fixture-unit rating (Fig. 19.11) of 4 or less and a maximum drain size of 2 in is above the battery, every vertical branch must have a continuous vent. If a fixture with a fixture-unit rating exceeding 4 is present, all fixtures in the battery must be individually vented.

Figure 19.11   Fixture-Unit Ratings in Zone 3

Bathtub	2
Shower	2
Residential toilet	4
Lavatory	1
Kitchen sink	2
Dishwasher	2
Clothes washer	3
Laundry tub	2

Figure 19.12   Vent Sizing for Zone 3

(For Use with Individual, Branch, and Circuit Vents for Horizontal Drainpipes)

Drainpipe size, in	Drainpipe grade, in/ft	Vent pipe size, in	Maximum developed length of vent pipe
$1\frac{1}{2}$	$\frac{1}{4}$	$1\frac{1}{4}$	Unlimited
$1\frac{1}{2}$	$\frac{1}{4}$	$1\frac{1}{2}$	Unlimited
2	$\frac{1}{4}$	$1\frac{1}{4}$	290 ft
2	$\frac{1}{4}$	$1\frac{1}{2}$	Unlimited
3	$\frac{1}{4}$	$1\frac{1}{2}$	97 ft
3	$\frac{1}{4}$	2	420 ft
3	$\frac{1}{4}$	3	Unlimited
4	$\frac{1}{4}$	2	98 ft
4	$\frac{1}{4}$	3	Unlimited
4	$\frac{1}{4}$	4	Unlimited

Circuit-vented batteries may not receive the drainage from fixtures on a higher level.

Circuit vents should rise vertically from the drainage. However, the vent can be taken off the drainage horizontally, if the vent is washed by a fixture with a rating of no more than 4 fixture units. The washing cannot come from a water closet. The pipe being washed must be at least as large as the horizontal drainage pipe it is venting (Fig. 19.12).

In zone 3, circuit vents may be used to vent up to eight fixtures utilizing a common horizontal drain. Circuit vents must be dry vents, and they should connect to the horizontal drain in front of the last fixture on the branch. The horizontal drain being circuit-vented must not have a grade of more than 1 in/ft. In zone 3, the horizontal section of drainage being circuit-vented is interpreted as a vent. If a circuit vent is venting a drain with more than four water closets attached to it, a relief vent must be installed in conjunction with the circuit vent.

All vents, except those for fixtures with integral traps, should connect above the trap's seal. A sanitary-tee fitting should be used going from a vertical stack vent to a trap. Other fittings, with a longer turn, such as a combination wye and eighth bend, will place the trap in greater danger of back-siphonage. I know this goes against the common sense of a smoother flow of water, but the sanitary tee reduces the risk of a vacuum.

The Main Water Pipe

The main water pipe delivering potable water to a building is called a *water service*. A water service pipe must have a diameter of at least $3/4$ in. The pipe must be sized, according to code requirements, to provide adequate water volume and pressure to the fixtures.

Ideally, a water service pipe should be run from the primary water source to the building in a private trench. By private trench, I mean a trench not used for any purpose, except the water service. However, it is normally allowed to place the water service in the same trench used by a sewer or building drain, when specific installation requirements are followed. The water service pipe must be separated from the drain pipe. The bottom of the water service pipe may not be closer than 12 in to the drain pipe at any point.

A shelf must be made in the trench to support the water service (Fig. 19.13). The shelf must be made solid and stable, at least 12 in above the drain pipe. It is not acceptable to have a water service located in an area where pollution is probable. A water service should never run through, above, or under a waste disposal system, such as a septic field.

If a water service is installed in an area subject to flooding, the pipe must be protected against flooding. Water services must also be protected against freezing. The depth of the water service will depend on the climate of the location. Check with the local code officer to see how deep a water service pipe must be buried. Care must be taken when backfilling a water service trench. The backfill material must be free of objects, such as sharp rocks, that may damage the pipe.

When a water service enters a building through or under the foundation, the pipe must be protected by a sleeve. This

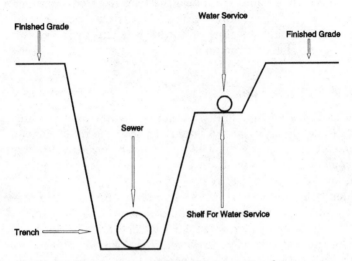

Figure 19.13 Water service and sewer in common trench.

sleeve is usually a pipe with a diameter at least twice that of the water service. Once through the foundation, the water service may need to be converted to an acceptable water distribution pipe. As you learned while reading about approved materials, some materials approved for water service piping are not approved for interior water distribution.

If the water service pipe is not an acceptable water distribution material, it must be converted to an approved material, generally within the first 5 ft of entry into the building. Once inside a building, the maze of hot and cold water pipes is referred to as water distribution pipes (Fig. 19.14).

Pressure-Reducing Valves

Pressure-reducing valves must be installed on water systems when the water pressure coming to the water distribution pipes is in excess of 80 pounds per square inch (psi). The only time this regulation is generally waived is when the water service is bringing water to a device requiring high pressure.

Figure 19.14 Common Minimum Fixture Supply Sizes

Fixture	Minimum supply size, in
Bathtub	$\frac{1}{2}$
Bidet	$\frac{3}{8}$
Shower	$\frac{1}{2}$
Toilet	$\frac{3}{8}$
Lavatory	$\frac{3}{8}$
Kitchen sink	$\frac{1}{2}$
Dishwasher	$\frac{1}{2}$
Laundry tub	$\frac{1}{2}$
Hose bib	$\frac{1}{2}$

Booster Pumps

Not all water sources are capable of providing optimum water pressure. When this is the case, a booster pump may be needed to increase water pressure. If water pressure fluctuates heavily, the water distribution system must be designed to operate on the minimum water pressure anticipated.

When calculating the water pressure needs of a system, you can use information provided by the code book. There are ratings for all common fixtures that show the minimum pressure requirements for each type of fixture. A water distribution system must be sized to operate satisfactorily under peak demands.

Booster pumps must be equipped with low-water cutoffs. These safety devices are required to prevent the possibility of a vacuum, which may cause back-siphonage.

Water Tanks

When booster pumps are not a desirable solution, water storage tanks are a possible alternative. Water storage tanks must be protected from contamination. They may not be located under soil or waste pipes. If the tank is a gravity type, it must be equipped with overflow provisions.

The water supply to a gravity-style water tank must be automatically controlled. This may be accomplished with a ball cock or other suitable and approved device. The incoming water should enter the tank by way of an air gap. The air gap should be at least 4 in above the overflow.

Water tanks must have provisions that allow them to be drained. The drainpipe must have a valve, to prevent draining, except when desired.

Pressurized water tanks are the type most commonly encountered in modern plumbing. These tanks are the type used with well systems. All pressurized water tanks should be equipped with a vacuum breaker. The vacuum breaker is installed on top of the tank, and it should have at least a $\frac{1}{2}$-in diameter. These vacuum breakers should be rated for proper operation up to maximum temperatures of 200°F.

It is also necessary to equip these tanks with pressure relief valves. These safety valves must be installed on the supply pipe that feeds the tank or on the tank itself. The relief valve discharges when pressure builds to a point of endangering the integrity of the pressure tank. The valve's discharge must be carried, by gravity, to a safe and approved disposal location. The piping carrying the discharge of a relief valve may not be connected directly to the sanitary drainage system.

Water Conservation

Water conservation continues to grow as a major concern. When the flow rates for various fixtures are set, water conservation is a factor. The flow rates of many fixtures must be limited to no more than 3 gallons per minute (gpm). In zone 3, these fixtures include the following:

Showers	Lavatories
Kitchen sinks	Other sinks

The rating of 3 gpm is based on a water pressure of 80 psi.

When installed in public facilities, lavatories must be

equipped with faucets producing no more than $\frac{1}{2}$ gpm. If the lavatory is equipped with a self-closing faucet, it may produce up to $\frac{1}{4}$ gpm per use. Water closets are restricted to a use of no more than 4 gal of water, and urinals must not exceed a usage of $1\frac{1}{2}$ gal of water, with each use.

Antiscald Precautions

It is easy for the very young or the elderly to suffer serious burns from plumbing fixtures. In an attempt to reduce accidental burns, mixed water to gang showers must be controlled by thermostatic means or by pressure-balanced valves. All showers, except for showers in residential dwellings in zones 1 and 2, must be equipped with pressure-balanced valves or thermostatic controls. These temperature control valves may not allow water with a temperature of more than 120°F to enter the bathing unit. In zone 3, the maximum water temperature is 110°F. In zone 3 these safety valves are required on all showers.

Valve Regulations

Gate valves and ball valves are examples of full-open valves, as required under valve regulations. These valves do not depend on rubber washers, and when they are opened to their maximum capacity, there is full flow through the pipe. Many locations along the water distribution installation require the installation of full-open valves. In zone 1, these types of valves are required in the following locations:

- On the water service, before and after the water meter
- On each water service for each building served
- On discharge pipes of water supply tanks, near the tank
- On the supply pipe to every water heater, near the heater

- On the main supply pipe to each dwelling

In zone 3, the mandatory locations of full-open valves are as follows:

- On the water service pipe, near the source connections
- On the main water distribution pipe, near the water service
- On water supplies to water heaters
- On water supplies to pressurized tanks, such as like well-system tanks
- On the building side of every water meter

In zone 2, full-open valves must be used in all water distribution locations, except as cutoffs for individual fixtures, in the immediate area of the fixtures. Other local regulations may apply to specific building uses; check with the local code officer to confirm where full-open valves are required in your system. All valves must be installed so that they are accessible.

Cutoffs

Cutoff valves do not have to be full-open valves. Stop-and-waste valves are an example of cutoff valves that are not full-open valves. Every sill cock must be equipped with an individual cutoff valve. Appliances and mechanical equipment that have water supplies are required to have cutoff valves installed in the service piping. Generally, with only a few exceptions, cutoffs are required on all plumbing fixtures. Check with the local code officer for fixtures not requiring cutoff valves. All valves installed must be accessible.

Backflow Prevention

Backflow and back-siphonage are genuine health concerns. When a backflow occurs, it can pollute entire water systems. Without backflow and back-siphonage protection, municipal water services could become contaminated. There are many ways to degrade the quality of potable water.

All potable water systems must be protected against back-siphonage and backflow with approved devices. Numerous types are available to provide this type of protection. The selection of devices is governed by the local plumbing inspector. It is necessary to choose the proper device for the use intended.

An air gap is the most positive form of protection from backflow. However, air gaps are not always feasible. Since air gaps cannot always be used, there are a number of devices available for the protection of potable water systems.

Some backflow preventers are equipped with vents. When these devices are used, the vents must not be installed so that they may become submerged. Also these units must be capable of performing their function under continuous pressure.

Some backflow preventers are designed to operate in a manner similar to an air gap. With these devices, when conditions arise that may cause a backflow, the devices open and create an open airspace between the two pipes connected to it. Reduced-pressure backflow preventers perform this action very well. Another type of backflow preventer is an atmospheric vent backflow preventer.

Vacuum breakers are frequently installed on water heaters, hose bibs, and sill cocks. They are also generally installed on the faucet spout of laundry tubs. These devices either mount on a pipe or screw onto a hose connection. Some sill cocks are equipped with factory-installed vacuum breakers. These devices open, when necessary, and break any siphonic action with the introduction of air.

In some specialized cases, a barometric loop is used to prevent back-siphonage. In zone 3, these loops must extend at least 35 ft high and can only be used as a vacuum breaker. These loops are effective because they rise higher than the point at which a vacuum suction can occur. Barometric loops work on the principle that since they are 35 ft high, suction is not achieved.

Double-check valves are used in some instances to control backflow. When used in this capacity, doublecheck valves must be equipped with approved vents. This type of protection would be used on a carbonated beverage dispenser, for example.

Backflow preventers must be inspected from time to time and so must be installed in accessible locations.

Some fixtures require an air gap as protection from backflow, such as lavatories, sinks, laundry tubs, bathtubs, and drinking fountains. This air gap is achieved through the design and installation of the faucet or spout serving these fixtures.

When vacuum breakers are installed, they must be installed at least 6 in above the flood-level rim of the fixture. Vacuum breakers, because of the way they are designed to introduce air into the potable water piping, may not be installed where they may suck in toxic vapors or fumes. For example, it is not acceptable to install a vacuum breaker under the exhaust hood of a kitchen range.

When potable water is connected to a boiler, for heating purposes, the potable water inlet should be equipped with a vented backflow preventer. If the boiler water contains chemicals, the potable water connection should be made with an air gap or a reduced-pressure backflow preventer.

Connections between a potable water supply and an automatic fire sprinkling system should be made with a check valve. If the potable water supply is being connected to a nonpotable water source, the connection should be made through a reduced-pressure backflow preventer.

Lawn sprinklers and irrigation systems must be installed with backflow prevention in mind. Vacuum breakers are a preferred method for backflow prevention, but other types of backflow preventers are allowed.

Hot Water Installations

When hot water pipe is installed, it is often expected to maintain the temperature of the hot water for a distance of up to 100 ft from the fixture it serves. If the distance between the hot water source and the fixture being served is more than 100 ft, a recirculating system is frequently required. When a recirculating system is not appropriate, other means may be used to maintain water temperature. These means could include insulation or heating tapes. Check with the local code officer for approved alternatives to a recirculating system, if necessary.

If a circulator pump is used on a recirculating line, the pump must be equipped with a cutoff switch. The switch may operate manually or automatically.

Water Heaters

The standard working pressure for a water heater is 125 psi. The maximum working pressure of a water heater is required to be permanently marked in an accessible location. Every water heater must have a drain, located at the lowest possible point on the water heater. Some exceptions may be allowed for very small, under-the-counter water heaters.

All water heaters must be insulated. The insulation factors are determined by the heat loss of the tank in 1 hour's time. These regulations are required of a water heater before it is approved for installation.

Relief valves are mandatory equipment on water heaters. These safety valves are designed to protect against excessive temperature and pressure. The most common type of safety valve used will protect against both temperature and pressure, from a single valve. The blow-off rating for these valves must not exceed 210°F and 150 psi. The pressure relief valve must not have a blow-off rating of more than the maximum working pressure of the water heater it serves; this is usually 125 psi.

When temperature and pressure relief valves are installed, their sensors should monitor the top 6 in of water in the water heater. No valves may be located between the water heater and the temperature and pressure relief valves.

The blow-off from relief valves must be piped down, to protect bystanders, in the event of a blow-off. The pipe used for this purpose must be rigid and capable of sustaining temperatures of up to 210°F. The discharge pipe must be the same size as the relief valve's discharge opening, and it must run, undiminished in size, to within 6 in of the floor. If a relief valve discharge pipe is piped into the sanitary drainage system, the connection must be through an indirect waste. The end of a discharge pipe may not be threaded, and no valves may be installed in the discharge pipe.

When the discharge from a relief valve could damage property or people, safety pans should be installed. These pans typically have a minimum depth of $1\frac{1}{2}$ in. Plastic pans are commonly used for electric water heaters, and metal pans are used for fuel-burning heaters. These pans must be large enough to accommodate the discharge flow from the relief valve.

The pan's drain may be piped to the outside of the building or to an indirect waste, where people and property will not be affected. The discharge location should be chosen so that it will be obvious to building occupants when a relief valve discharges. Traps should not be installed on the discharge piping from safety pans.

Water heaters must be equipped with an adjustable temperature control. This control must be automatically adjustable from the lowest to the highest temperatures allowed. Some locations restrict the maximum water temperature in residences to 120°F. There must be a switch to shut off the power to electric water heaters. When the water heater uses a fuel, such as gas, there must be a valve to cut off the fuel source. Both the electric and the fuel shutoffs must be able to be used without affecting the remainder of the building's power or fuel. All water heaters requiring venting must be vented in compliance with local code requirements.

Common Sense

Common sense plays a big role in the successful design of a plumbing system. If you know the local plumbing code, you should be able to design residential systems with ease. To create designs that are cost-effective, use common sense.

Lay out your jobs on paper (Fig. 19.15), and then look at what you've done. Could you combine some vents that are shown as individual vents? Would it be better to run all of a

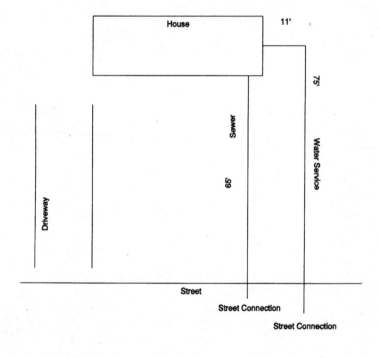

Figure 19.15   Site plan with city water and sewer.

bathroom's water pipe in the partition walls, or would it be easier and cheaper to bring the pipes up under each fixture? Just study your initial design and look for ways to improve it. I'll bet you'll find some.

INDEX

Abbreviations, 330
Aerators, clogged, 33
Air break, 400
Air-conditioner waste, 402
Air gap, 19, 26, 37, 39, 137, 400–401, 417
Air tools, safety with, 309
Antiscald devices, 4, 415
Appliances, plumbing, 269
Approved materials, 242
Atmospheric pressure, 336

Backflow, protection against, 18, 20, 138, 416–418
Bar sink faucets, 112
Barometric loop, 417
Basement bathroom, 222
Bathing units for the handicapped, 127
Bathing units, waterproof walls for, 139
Bathroom, basement, 222
Bathtub faucets, 112
Bathtubs, installation of, 134
  troubleshooting, 147
Bleed fittings, 43
Blowout toilet, 126
Branch-interval detail, 405, 407

Branches, horizontal, 367
Brass pipe, 11
Building drains, problems with, 97
  sizing, 365

Chamber-type septic systems, 199
Check valves, 20
Chemical wastes, 402
Chisels, safety with, 311
Chlorinated polyvinyl chloride pipe, 5, 6, 12
Circle, area of, 342
  circumference of, 341
  formulas for, 338
Circuit vents, 73
Clear-water wastes, 400
Closet augers, 237
Clothing safety, 301
Code considerations, 55
Commercial fixtures, 142
Commercial showers, 143
Common vents, 71
Compression leaks, 50
Conduit, 289
Connections, materials, 264–268
Conversion charts, 329
Conversion factors for measurements, 339

Conversion tables, 329, 340
Converting inches to decimals of
    feet, 335
Copper pipe, 5, 9, 11
    leaks in, 40
    steam in, 42
    water trapped in, 43–44
Coworkers, safety among, 312–314
Critical level, 21
Cross connections, 137
Cubes of numbers, 346–347
Customer facilities, fixtures
    required for, 122

Dangerous wastes, 402
Day care facilities, fixtures
    required for, 121
Decimals converted to millimeters,
    350
Decimal equivalents of fractions of
    an inch, 337, 353–354
Decimals equivalents of an inch,
    351
Design criteria, 393
Design temperature, 331
Designing plumbing systems, 393
Dishwashers, 278, 292
    dangers of, 137
    electrical connections for, 292
    installation of, 278–279
    troubleshooting, 279–282
Disposable faucets, 105
Ditches, safety in and around, 313
Diverters, problems with, 113–114
Drainage problems, related to
    materials, 96
Drainage systems:
    cast-iron, 87
    copper, 86
    galvanized steel, 87, 218
    plastic, 86
Drainage unit ratings, in zone 3,
    362
Drain pipes, problems with, 93

Drains, bathtubs, 88
    clogged, 233
    floor, 138
    overloaded, 95
    problems with, 93
    slow-moving, 86
    shower, 90
    sinks, 90
    toilets, 91
Drain cleaners, safety with,
    307–308
Drain openers:
    air-powered, 236–237
    large, 238–239
    liquid and powdered, 235–236
    supersize, 239
    water-powered, 236
Drain-waste-and-vent (DWV) sys-
    tems:
    installing, 53, 59, 65, 67
    materials approved for, 250–256
    repairing, 85
    riser diagram of, 404
    troubleshooting, 85
Drills, safety with, 306–307
Drinking fountains, 142
    for the handicapped, 129
Drips, from faucet spout, 106
Dry vents, 69

Ear protection, 301
Electrical basics, 285
    circuits, high-voltage, 219
    color codes, 285
    connections, 290
    wire nuts, 286
Electrical boxes, attachment of,
    288
    rough-in numbers, 288
    selection of, 286–288
Electrical connections, 290
    for dishwashers, 292
    for garbage disposers, 292
    for water heaters, 293

Electrical outlets, in floors, 291–292
Electrical wires, snaking, 289–290
Employee facilities, fixtures required for, 122
Eye protection, 301
Expansion in plastic pipe, 331

Faucet:
  drips in, 106
  grit in, 106
  leaks around base of, 108
  leaks around handles of, 108
  leaks from below, 111
  won't turn off, 108
Faucet handles, leaks around, 113
Faucet seats, 107
Faucet spray, limited water from, 111
  no water from, 110
Faucet washers, 104, 107
Faucets, 103
  ball-type, 105
  bar sink, 112
  bathtub, 114
  cartridge-style, 104
  disposable, 105
  for the handicapped, 127
  kitchen, 103
  laundry, 112
  lavatory, 111
  shower, 112
  single-handle,107
  wall-mounted, 114
  washerless, 104, 107
Field conditions, 115
Filters, water, 33
  clogged, 33
First aid, 315
  for back injuries, 322
  for bleeding, 316–319, 321
  for blisters, 322
  for burns, 324–326
  for cramps, 326

First aid (Cont.):
  for eyes, 320
  for exhaustion, 326–327
  for facial injuries, 321
  for hand injuries, 322
  for infections, 319
  for leg injuries, 322
  for nosebleeds, 321
  for open wounds, 316
  for scalp injuries, 320
  for shock, 323–324
  for splinters, 319
  tourniquets, 318
Fittings, approved, 261
  bleed, 43
  drain, 43
  vent, 43
Fixture placement, 80, 130
Fixture-unit rating, in zone 3, 362
  values, 387
Fixture-supply minimums, 383
Fixture-supply outlet capacities, 384
Fixtures, 119
  handicap, 123–126
  required, 119
  required for customer facilities, 122
  required for day care facilities, 121
  required for employee facilities, 122
  required for multifamily buildings, 120
  required for nightclubs, 120
  required for single-family residence, 120
Fixtures, access of, 135
  approved, 263
  floor mounted, 133
  installation of residential, 136
  sealing, 132
  securing, 132
Fixtures, commercial, 142
Fixtures, special, 143

Flanges, approved, 262
Flood-level rim, 73
Floor-mounted fixtures, 133
Flow-rate equivalents, 331, 350
Flow velocity, 384
Flush holes, in a toilet, 91
Flush valves, 144
Formulas, 334
Fractions converted to decimals,
    352
Frost-free silcocks, 114
Frozen drains, 94
    vents, 100
    water pipes, 50

Galvanized steel pipe, 11, 32
Garbage can washers, 143
Garbage disposers, 269, 292
    electrical connections for, 292
    commercial, 143
    troubleshooting, 155, 269–275
Gas, septic, 208
    sewer, 99
Gate valve, 162
Grab bars, for the handicapped,
    129
Grade requirements for pipes, 66,
    394
Grinders, safety with, 312
Ground fault interrupters, 291
Gutters, sizing, 373

Hand-held showers, 137
Handicap bathing units, 127
Handicap drinking fountains, 129
Handicap faucets, 127
Handicap fixtures, 123–127
    design of, 126
    required locations, 123
    spacing for, 124
Handicap grab bars, 128
Handicap lavatories, 127
Handicap sinks, 127

Hard water, 33
Hole sizes, 66
Horizontal pipe supports, 81
Hose bibbs, 114
Hot water, 158

Ice makers, 275
    connections for, 275
    troubleshooting, 276–277
Indirect wastes, 399
    trapping of, 400
Individual vents, 72
Irrigation systems, 418
Island vents, 78–79

Jet pumps, 174
    installation of, 174–177
    troubleshooting, 181–188
Jewelry safety, 301

Kitchen faucets, 103
Kitchen sinks, troubleshooting, 154
Knee pads, 302

Ladders, safety with, 309–310
Laundry faucets, 112
Laundry trays, minimum drain
    size of, 138
    troubleshooting, 146
Lavatory faucets, 111
Lavatories, handicapped, 127
    minimum drain size of, 138
    mounting of, 134
    troubleshooting, 149
Lead pots and ladles, safety with,
    305
Leaks, in brass pipe, 48
    in copper pipe, 40
    in polybutylene pipe, 46
    in polyethylene pipe, 46
Liquid measures, 332

Low water pressure, 30, 109, 113

Manifold systems, 27
Material mistakes, 96
Materials, approved, 242
  for DWV systems, 250–256
  for water distribution, 248–250
  for water services, 244–248
Material connections, 268
Materials, selection of, 241
Metric conversions, 332–333,
    359–360
Mound-type septic systems, 202
Moving water, 45
Multifamily buildings, fixtures
    required for, 120
Multipliers, 343

Nail plates, 76
Nightclubs, fixtures required for,
    120
Nipples, approved, 262

Offsets, in horizontal piping, 398
  fittings approved for, 398
Offsets from horizontal to vertical
    runs, 399
  fittings approved for, 399
Offsets in vertical piping to hori-
    zontal runs, 399
  fittings approved for, 399
Old plumbing fixtures, 218
OSHA, 298
Overloaded drains, 95

Pans, retaining, 160
  shower, 140
Pipe:
  brass, 11
  CPVC, 5, 6, 12
  copper, 5, 9, 11

Pipe (Cont.):
  galvanized steel, 11, 32
  polybutylene, 5, 7, 12
  undersized, 33
  water service, 244–248
Pipe penetrations, 268
Pipe sizing, 361
  example, 366
  fixture-unit method, 387
  rule-of-thumb method, 386
Pipe size reduction, 397
Pipes, squeaking, 38
  unsecured, 37
  water, 5, 10
Pipe support intervals, 394–396
Pipe threaders, safety with, 308
Pitch requirements for pipes, 66,
    394
Plugs, approved, 262
Plumbing, above-ground, 65
  dangers of, 295
  underground, 28, 54, 56,
Plumbing, moving old, 219–220,
    225–234
Plumbing systems, designing, 393
Pools, swimming, 402
Polybutylene pipe, 5, 7, 12
Potable water, 1
Power-actuated tools, safety with,
    309
Pressure tanks, 30
Pressure-relief valve, 158, 163,
    419
Product knowledge, 117
Protective pipe plates, 76
Pumps, booster, 413
  deep-well jet, 177–178
  jet, 174
  recirculating, 161
  sewage, 381
  submersible, 178

Rain leaders, sizing, 373
Recirculating pumps, 161, 418

Reduced-pressure backflow preventer, 417
Regulations, toilets, 141
Relief vents, 71, 77
Remodeling, 215
    old plumbing fixtures, 217–218, 223–232
    pipe sizing during, 215–216
Removing old plumbing fixtures, 218
Residential fixtures, installation of, 136
Retaining pan, 160, 420
Reverse trap toilet, 126
Riser diagram, of DWV system, 404
    of water system, 22–23
Rotted walls, 220–221
Roof drains, sizing, 373
Rough-in measurements, 25

Safety, 295
    general, 299
Saws, safety with, 307
Scaffolds, safety with, 310
Screwdrivers, safety with, 311
Septic gases, 206
Septic systems, 195, 219
    chamber type, 199
    components of, 196
    function of, 203–204
    mound type, 202
    trench type, 200
    troubleshooting, 207–213
Septic tanks, types of, 198
    maintenance of, 204–206
Sewage pump, 207
Sewer gas, 99
Sewers, common trench, 411
Sizing, 365
Shower bases, minimum size of, 139
Shower drains, minimum size of, 139
Shower pans, 140

Showers, commercial, 143
    gang, 5
    hand-held, 137
    installation of, 135
Shower head, drips from, 113
    flow from, 4
    securing, 138
Shower faucets, 112
    won't cut off, 113
Silcocks, frost-free, 114, 145
    troubleshooting, 145
Single-family residence, fixtures required for, 120
Single-handle faucets, 107
Sinks, for the handicapped, 127
Sinks, minimum drain sizes for, 141
Siphon jet toilet, 126
Siphon vortex toilet, 126
Siphon wash toilet, 126
Site plan, 421
Snakes, flat-tape, 237
    electric, 238
    manual, 238
Special fixtures, 143
Special wastes, 402
Sprinklers, lawn, 418
    fire, 418
Stack vents, 70
Stacks, cutting cast-iron, 221–222
Stacks, sizing, 368–371
Sterilizers, 402
Storm-water drainage, approved materials, 256–261
    sizing, 371–372
Square measure, common, 357
    other, 358
Square roots of numbers, 344–345
Squares of numbers, 348–349
Squeaking pipes, 38
Standpipe, washing machine, 79, 400
States in zone 1, 3
    in zone 2, 3
    in zone 3, 4

Stop-and-waste valve, 163
Submersible pumps, installation of, 178–181
  troubleshooting, 188–193
Sump vents, 381
Support intervals of fittings, 397
Support intervals of pipe, 81, 82, 394–397
Swimming pools, 402

Tanks, pressure, 30
  water, 413
Temperature-and-pressure-relief valves, 158, 163
Temperature conversions, 331
Toilet, minimum width requirement, 131
  proper alignment of, 133
  regulations for a, 141
Toilets, blowout, 126
  reverse trap, 126
  siphon jet, 126
  siphon vortex, 126
  siphon wash, 126
  tank parts of, 92
  troubleshooting, 35
Toilets, installation of, 135
  commercial installation of, 144
  troubleshooting, 150
Tool safety, 302–304
Torch safety, 304–305
Trap, sizes, 362
  zone, 1, 363
  zone 2, 363
  zone 3, 364
Traps, sand in, 96
Trap-to-vent distances, 382–383
Trench-type septic systems, 200
Troubleshooting, bathtubs, 147
  dishwashers, 279–282
  DWV systems, 85
  garbage disposers, 159, 269–275
  ice makers, 276–277
  kitchen sinks, 154

lavatories, 149
laundry tubs, 146
showers, 148
silcocks, 145
toilets, 35, 150
washing machine hook-ups, 146, 283–284
water heaters, 164–167
water systems, 29
Troubleshooting pumps, 181
  jet, 181–188
  submersible, 188–193
Troubleshooting septic systems, 207–213
Tub wastes, adjustment of, 89
Tubs, whirlpool, installation of, 144

Underground plumbing, 54, 56
Underground water pipes, 28
Undersized piping, 33
Unsecured water pipes, 37
Urinals, commercial installation of, 144

Vacuum breakers, 20, 24, 417
Valve regulations, 415–416
Valves, 17, 31
  check, 20
  double-check, 417
  flush, 144
  gate, 162
  pressure-relief, 158, 163
  saddle, 230
  stop-and-waste, 163
Valves, approved, 262
Vehicle safety, 299–300
Vent, graded connection of, 80, 408
  penetration, 78, 83
Vent stacks, 378
Venting, offsets, 407–408
Venting batteries of fixtures, 409
Vents, branch, 76
  circuit, 73, 409

Vents, branch (*Cont.*):
    common, 71
  crown, 381
  dry, 69, 408
  individual, 72
  island, 78–79
  relief, 71, 77, 409
  stack, 70, 378, 409
  wet, 68
  yoke, 76
Vents, clogged, 100
  frozen, 100
Vents, flashing, 403
Vents, main, 403
Vents, problems with, 99
Vents, protection of, 405
Vents, sidewall, 406
Vents, sizing, 374–381
Vertical pipe supports, 82

Wall-mounted faucet, 114
Walls, rotted, 220–221
Washers, faucet, 104, 107
Washerless faucets, 104, 107
Washing machine standpipe, 79,
    400
Washing machine hook-ups, trou-
    bleshooting, 146, 281–282
Washing machines, 281
Wastes, air conditioner, 402
  chemical, 402
  clear-water, 400
  dangerous, 402
  indirect, 399
  special, 402
  trapping of, 400
Water conservation, 1, 414
Water coolers, 142
Water distribution system, 1
  installing, 16
  materials approved for, 248–250

Water distribution system (*Cont.*):
  repairing, 29
  riser diagram of, 22–23
  sizing, 1, 12, 382
  troubleshooting, 29
Water, feet heat to PSI, 355–357
Water flow, nonuniform, 110, 114
Water hammer, 34, 39
Water, hard, 33
  moving, 45
Water heaters, 157, 419
Water heaters, electric, 161, 163,
    217, 291
  electrical connections for, 291
  troubleshooting, 164–167
Water heaters, gas-fired, 167, 217
  troubleshooting, 167–170
Water heaters, oil-fired, 170, 217
  troubleshooting, 170–171
Water, hot, 158
Water pipes, main, 411
  offsets in, 40
  underground, 28
Water pressure, excessive, 34
  low, 30, 109, 113
Water service pipe, 244
  in common trench, 411
  materials for, 244–248
  protection of, 412
Water temperature, maximum, 142
Water, volume to weight, 358
Waterproof walls, 139
Wax ring, for a toilet, 92
Well systems, 173
Wet vents, 68
Whirlpools tubs, installation of,
    141
Wire molding, 289
Wire nuts, 286

Yoke vents, 76

ABOUT THE AUTHOR

R. Dodge Woodson is a general contractor, plumbing contractor, and licensed master plumber with over 20 years' experience in both residential and commercial work. His previous books include *National Plumbing Codes Handbook, The Plumbing Apprentice Handbook, The Plumber's Troubleshooting Guide*, and *Plumber's Quick Reference Manual*. He lives in Bowdoinham, Maine.